BHB

# Solid State Devices 1983

# Solid State Devices 1983

Ten invited papers presented at the Thirteenth European Solid
State Device Research Conference (ESSDERC) and the Eighth
Symposium on Solid State Device Technology (SSSDT) held at
the University of Kent at Canterbury, 13–16 September 1983

Edited by E H Rhoderick

PHYS
Sep/ae

Conference Series Number 69

The Institute of Physics (Great Britain))
Bristol and London

CODEN IPHSAC 69 1-180 (1984)

*British Library Cataloguing in Publication Data*

European Solid State Device Research
  Conference (*13th: 1983: Canterbury*)
  Solid state devices 1983. — (Conference series/
  Institute of Physics, ISSN 0305-2346; no. 69)
  1. Semiconductors — Congresses
  I. Title      II. Rhoderick, E H
  III. Symposium on Solid State Device Technology
  (*8th: 1983: Canterbury*)
  iv. Series
  621.3815'2      TK7871.85      *6964-5516*
                                  *PHYS*

ISBN 0-85498-160-8
ISSN 0305-2346

Organising Committee
    J C Walling (Chairman), K J S Cave (Co-Chairman), B L H Wilson (Vice-Chairman),
    D H Paxman (Secretary), C Hilsum, R J Bennett (Local Arrangements)

Honorary Editor
    E H Rhoderick

The 13th European Solid State Device Research Conference and the 8th Symposium on Solid
State Device Technology were organised by The Institute of Physics and co-sponsored by The
European Physical Society, The Dutch Physical Society, The German Physical Society, The
Institution of Electrical Engineers, The Institution of Electronic and Radio Engineers, The
Institute of Electrical and Electronics Engineers (United Kingdom and Republic of Ireland
Section).

Published by The Institute of Physics, Techno House, Redcliffe Way, Bristol BS1 6NX, and
47 Belgrave Square, London SW1X 8QX, England.

Printed in Great Britain by J W Arrowsmith Ltd, Bristol.

# Preface

Following the now established pattern, the 13th European Solid State Device Research Conference and the 8th Symposium on Solid State Device Technology were held concurrently from 13–16 September 1983, at the University of Kent at Canterbury. This ostensibly cumbersome arrangement has evolved historically and, like many apparent anomalies, is justified by the fact that it works. The boundary between device research and technology grows more indistinct every year, but the possibility of unfilled gaps or unnecessary duplication occurring appears to have been successfully minimised by close cooperation between the two programme committees.

This year the papers were fairly evenly balanced between the Conference and the Symposium. In the Conference, over half of the 46 papers were concerned with MOSFETs or MESFETs, compared with a mere 8 devoted to bipolar devices. This imbalance probably reflects a greater number of unsolved problems relating to FETs rather than their importance *vis-à-vis* bipolars. The large number of papers (14) devoted to computer simulation (modelling) was also noteworthy; there is a possible danger here in that people with good computer facilities at their disposal may be attracted into this field without having the necessary experimental back-up to make their efforts bear fruit. There was a mere handful of papers (4) concerned with novel devices, most of these being variants of existing ones.

In the Symposium, over half of the 51 papers were devoted to silicon technology, as one would expect, with the III–V compounds accounting for only 9. The remainder were distributed between insulator and contact technology and failure mechanisms.

As is the normal practice, the Proceedings contain only the invited review papers. These to a certain extent make up for the lack of contributed papers on novel devices, with stimulating accounts of applications of superlattices, hot electron transistors and liquid-crystal optical processing. The remaining invited papers are all in the field of technology, and confirm the main theme of the Conference/Symposium as consolidation rather than innovation.

**September 1983**                                                     **E H Rhoderick**

# Contents

*Inst. Phys. Conf. Ser. No. 69*
*Paper presented at ESSDERC/SSSDT 1983, Canterbury 13–16 Sept. 1983*

# Quantum well structures and superlattices and their applications potential

Arthur C. Gossard

Bell Laboratories, Murray Hill, New Jersey 07974 USA

Abstract. Epitaxial crystal growth techniques now provide much flexibility in design and construction of multiple thin-layer semiconductor
structures. Quantum confinement, quantum tunnelling and two-
dimensionality come into force in determining electron behavior in
these materials and lead to fundamental new behaviors. The potential
of present and emerging devices based on these phenomena is assessed.
Optical applications include the quantum well laser, avalanche photodiode, superlattice infrared detector and superlattice optical bistable
element. Electrical applications based on modulation doping, on
tunnelling and on compositional grading of multiple layer structures
are also considered.

## 1. Introduction

Epitaxial crystal growth techniques are now available which can synthesize finely layered single-crystal films with layer thicknesses down to the scale of atomic monolayers. In the attainable size range, particles are subject to quantum confinement and tunnelling effects. Interest in such substances arises from the new properties which occur when the structures are small enough to produce particle energies and wavefunctions which are substantially modified relative to bulk or unconfined material. This in turn produces a spectrum of new physical properties. This paper reviews a number of these new properties and their possible application. For the cases in which the particles of interest are electrons or holes, the differences in electron affinity between adjacent layers can easily exceed the kinetic energy of conduction electrons, permitting the confinement of quantized electron gases in the layers. Materials in which electrons are confined in quantum states in thin layers are referred to as quantum well structures. Superlattices are multiple-period structures displaying quantum confinement or tunnelling. Barrier tunnelling or penetration is not a necessary condition for new quantum effects to be produced, though, and particles in adjacent layers may be coupled or independent.

When the particles being considered are photons or phonons, they will in general not be bound in individual layers, but will nonetheless interact with superlattices in an important way. One of the first applications of artificially layered materials, in fact, was the use of Langmuir-Blodgett organic multilayer films to diffract long wavelength x-rays. Production of artificial superlattices for x-ray diffraction mirrors has continued and been highly developed. A large amount of other work on metallic multilayers is being pursued, although this is not the main topic of this paper. The metal multilayers are a fruitful source of information on film and crystal growth processes and on the occurrence of ordering in alloys. They show structure-related elastic anomalies, structure-dependent magnetic properties, new superconducting behavior, and unique melting behavior. Much further development can be anticipated for metal multilayer systems. Here, however, we will concentrate on developments obtained with semiconductor systems, where more perfect structures have been formed, where new quantum effects have been more clearly observed, and where the greatest applications potential is currently being realized.

In Section 2 of this paper, we discuss growth techniques of multilayer and quantum well semiconductors. Section 3 presents the quantization of electrons and related optical properties of these materials and the application to quantum well light source and optical logic elements. Section 4 describes transport across barrier layers, with application to light detectors and to unipolar rectification and amplification. Modulated doping and the resultant high mobility electron transport along layers are treated in Section 5, along with high speed devices and new two-dimensional effects in the structures. Section 6 concerns additional areas of research which are currently developing in this field.

## 2. Growth Techniques

Semiconductor heterostructures can be grown by several methods. The first major application of heterostructures was to semiconductor lasers, where the heterostructures provided both confinement of electrons and confinement and guiding of light (Kroemer 1963; Hayashi and Panish 1969). Electron confinement was produced by the difference in electron affinity and light confinement by the difference in index of refraction between adjacent heterolayers. The semiconductor layers were grown by liquid phase epitaxy of GaAs and (Al,Ga)As and the minimum thicknesses of layers which were required were 100 to 200 nm. Control of thickness in LPE growth has been refined, and multiple-layer lasing structures with (In,Ga) (P,As) layers as thin as 15 nm, separated by InP layers have been reported (Rezek et al 1980).

Metal-organic chemical vapor deposition (MOCVD) has more recently been developed and applied in growth of thin-film multilayer semiconductor structures. Multiple layers down to a few nanometers thickness per layer have been produced and applied to lasers (Coleman et al 1981). The MOCVD technique is successful with mixed compound layer growth, although inhomogeneities or pinholes in the mixed layers have been reported (Holonyak 1980). Chemical vapor deposition without use of the metal-organic reactants is less useful for multiple layer growth because of its limited ability to grow aluminum compounds and has thus far not produced equivalently sharp interfaces.

Molecular beam epitaxy (MBE), in which atomic and molecular fluxes are created by evaporation or sublimation from elemental sources, is the most highly developed technique for multiple and thin-layer crystal growth. Beams of evaporating fluxes can be turned on and off with shutters in times less than required for growth of an atomic monolayer. Under properly chosen growth conditions, surface roughness can be smoothed to nearly monolayer scale by virtue of surface migration of atoms during growth to lower spots on surfaces. Composition modulations down to alternate single atomic layer thicknesses can be produced with MBE (Gossard 1976). Total film thicknesses may contain thousands of the thinnest layers. The most thoroughly examined materials have been the (Al,Ga)As/GaAs layered systems.

## 3. Quantum Levels and Optical Phenomena

The quantized energy levels of thin semiconductor layers have been extensively probed by optical techniques (Dingle 1975; R C Miller et al 1980; Pinczuk and Worlock 1982). The levels are shifted to higher energy by the electron and hole quantization in the GaAs layers. In the simplest approximation, the electrons and holes may be taken to have bulk effective masses and to be confined by square-well potentials derived from the heterojunction band discontinuities. This approximation accounts semiquantitatively for the observed spectra in GaAs quantum wells (Figure 1) if $\approx 85\%$ of the bandgap discontinuity occurs in the conduction band and $\approx 15\%$ in the valence band (Figure 2). For the n-th confined state, which contains n electron half-wavelengths in a well thickness L, the quantum energy is $h^2 n^2 / 8m^* L^2$ for the case infinitely high barriers and less with finite barriers. The most accurate representation of the transition energies requires a band structure calculation of the states (Schulman and McGill 1981).

### 3.1 Luminescence and Lasers

Quantum well luminescence occurs when electrons and holes bound in quantum well states radiatively recombine. This radiation is shifted to higher energy relative to bulk luminescence by the thickness dependent electron and hole quantum confinement energies. While the dominant luminescence from bulk material at low power generally comes from electron to impurity recombination, the GaAs quantum well luminescence is dominated by intrinsic electron-hole excitonic recombination, (Weisbuch et al 1981). Furthermore, the intensity of the GaAs quantum well luminescence is generally enhanced relative to that from thick layers of GaAs grown under comparable conditions (Petroff et al 1981). And

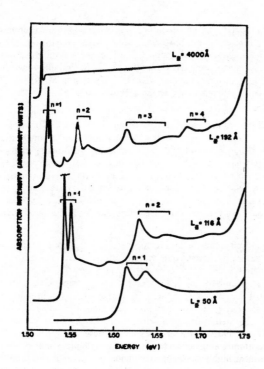

Figure 1. Optical absorption spectra of GaAs multiple quantum well structures of quantum well thickness $L_z$ at low temperature.

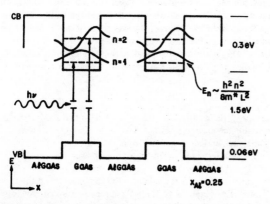

Figure 2. Conduction and valence band potential profiles and confined quantum states for GaAs/(Al,Ga)As quantum well structure.

finally, the luminescence spatial distribution from the quantum well structures as studied by high resolution scanning cathodoluminescence is more uniform than conventional thick-layer luminescence (Petroff et al 1981). All of these properties provide an attractive environment for laser operation, even though one might have initially feared that quantum well barriers would degrade or at least dilute luminescence performance. In fact, a variety of quantum well lasers with good operating characteristics have been made. Modified multi-quantum-well lasers in which carriers are confined in a smaller volume than the light wave have shown current thresholds below 30 mA (Tsang, 1981). Lifetime tests of quantum well lasers are also yielding promising results. Conventional MQW lasers operating at 3mW/mirror output power have shown median lifetimes of $\approx$ 5000 hours at 70°C (W. T. Tsang 1982). Low degradation rates are also being found in MOCVD-grown quantum well lasers (R. D. Dupuis et al 1983). In the high-power shorter wavelength visible regime, quantum well lasers also hold promise. By choice of a small quantum well thickness, the emission energy can be raised to the visible. Single quantum well lasers have produced up to 100 mW of cw visible radiation at room temperature

(Burnham et al 1983). In the longer wavelength range, quantum well lasers also hold promise. $In_{0.47}Ga_{0.53}As$ is a favorable longer wavelength laser material in that it can be grown on InP substrates to which it is lattice matched. Its bandgap, however, is about 0.1 eV below an optimum fiber transmission window at $1.33\mu$ wavelength. Its lasing energy has recently been raised to the fiber window value by incorporation in an $In_{0.47}Ga_{0.53}As/In_{0.47}Al_{0.53}As$ quantum well laser (Temkin et al 1983). The Ga-containing and Al-containing layers are both lattice-matched to the InP substrate.

Other systems which may be promising for long wavelength applications are lattice-mismatched superlattices such as GaP/GaAs and InAs/GaAs and superlattices based on HgTe and CdTe. In spite of mismatches, the lattice-mismatched superlattices can form well ordered strained-layer superlattices (Osbourn et al 1982) and give laser action (Ludowise et al 1983). Superlattice zone-folding can give weakly direct components to otherwise indirect-gap materials. Unfortunately, at high excitation levels, many of the mismatched, strained-layer heterostructures develop dislocation networks and are not stable (Ludowise et al 1983). The GaSb/AlSb system has band gaps in the technically interesting 1.3 $\mu$m system but has indirect gaps that lie very near the direct gap. The CdTe/HgTe superlattice encompasses band gaps which are direct from 1.6 eV to zero and are interesting for far infrared sources and detectors.

The sources of the improved efficiency, uniformity and intrinsic characteristics of quantum well lasers relative to thicker-film laser structures have not yet been completely delineated. Several factors are clearly important, though. One is the spatial confinement of electron-hole pairs into layers, which restricts possible migration to radiationless recombination centers at defects or impurities. Another factor entering enhanced GaAs quantum well laser behavior is interface quality. It has been noted that thick ($\geq 1000$Å) (Al,Ga)As layers often show rougher and less pure interfaces than do comparable thinner layers. This implies that GaAs double heterostructures with conventionally thick ($>1000$Å) (Al,Ga)As cladding layers have (Al,Ga)As interfaces which are inferior to those in multiple quantum wells. A final reason for the improved quantum well laser characteristics is a fundamental difference between the density of electron energy states near the band edges of the quantum well structures and thicker layers. The density of energy states for the two dimensional electrons and holes in quantum wells has a step-function form with constant density of states $dn/dE = m^*/\pi\hbar^2$ above the band edge (up to the energy of the next excited quantum state). In three dimensions, the density of states is proportional to the square root of energy above the band edge. The difference between these two forms of the density of states concentrates more electron states at energies near the band edge for quantum wells. This makes more electrons available for lasing in the quantum wells and leads to lower quantum well laser threshold (Dingle and Henry 1976).

### 3.2 Optical Bistability

Substances with strong easily saturable features in their optical transmission spectra are good candidates for optical switches or logic elements. Semiconductor exciton lines can have such characteristics, and the bulk GaAs exciton line was the source of the first optical bistability in a solid at a practicable power level (Gibbs et al 1979). Low temperature bistability in transmitted optical power was produced by saturation of the bulk GaAs exciton, which in turn changed the index of refraction of the GaAs and detuned an optical cavity formed by mirrors on the faces of a three-layer sandwich containing the GaAs. Room-temperature optical bistability was difficult to obtain with the bulk material, though, because of the decrease in amplitude of the exciton absorption of bulk GaAs under thermal dissociation of the exciton at room temperature.

In quantum wells, on the other hand, excitons are bound more strongly because of the two-dimensional confinement of electrons and holes (R C Miller et al 1981), and excitonic absorption peaks can remain strong and easily saturable at room temperature (Figure 3), (D A B Miller et al 1982). This permitted the first low-power observation of optical bistability at room temperature. The quantum well bistability measurements were performed on Fabry-Perot etalons formed by a GaAs/(Al,Ga)As superlattice, cladding layers, and reflective coatings. Powers of approximately $10kW/cm^2$ over a $10^{-6}cm^2$ spot were required to produce switching (Gibbs et al 1982).

Optically-pumped sources of spin-polarized electrons are yet another interesting application of quantum wells. The splitting of the optical absorption peaks in quantum well structures is produced by the shift of light and heavy hole states in the valence bands. With the hole states split by the quantum well

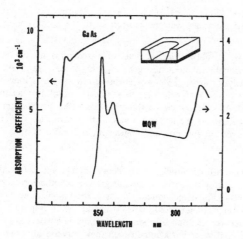

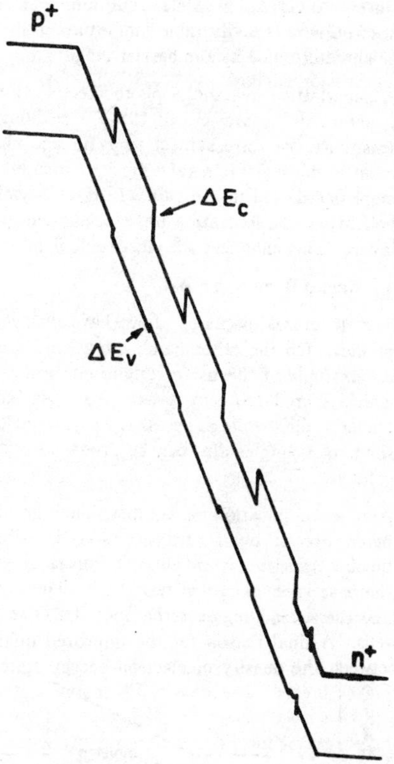

Figure 3. Exciton absorption edge for bulk GaAs and multiple quantum well structure at room temperature.

Figure 4. Band edge profiles under reverse bias for p-i-n superlattice avalanche photodetector.

potential, it becomes possible to optically pump electrons from the heavy hole band alone, resulting in a higher spin polarization than for the bulk case where both hole states are pumped and produce cancelling contribution to spin polarization (R C Miller et al 1979).

## 4. Barriers

The potential barriers between semiconductor layers are typically less than the fundamental band gaps of the semiconductors and thus of order tenths of electron volts, although in special cases, such as the combination GaSb/InAs, the potential barrier can slightly exceed the bandgaps. Electrons or holes can penetrate barriers of these heights substantially so that even the confined electron states have appreciable wavefunction amplitudes within the edges of the barriers. Quantum electron levels confined between pairs of thin barriers can be detected by resonant tunnelling under bias through the barriers (Chang et al 1974). Quantum wells separated by thin ($\approx 10\text{Å}$) barriers, are coupled sufficiently by tunnelling that their electron states are appreciably shifted and pairs of wells show well resolved bonding and antibonding combinations of electron levels (Dingle et al 1975). With thin barriers, properties are sensitive to alloy clustering in the barriers and to barrier thickness variations, an effect which has been noted particularly with MOCVD growth of mixed compound (Al,Ga)As barriers (Holonyak et al 1980). When many layers are present, the coupling between layers leads to the formation of minibands. These are bands formed from the quantum bound states of the individual layers, and the width of the minibands increases as barrier thicknesses and heights are made smaller. An early proposal for application of miniband formation involved electron reflection from minizone boundaries (Esaki et al 1970), leading to negative resistance and to Bloch oscillations of the current. Neither effect has been clearly established, presumably because electron scattering rates have thus far been too rapid to permit electron excitation to the zone boundaries by collisionless travel through a sufficient number of periods.

Transport of carriers across barrier layers can be assisted by free-carrier photoexcitation to energies above the barrier height. This has recently been employed in an infrared detector based on a

GaAs/(Al,Ga)As multiple quantum well structure (Chiu et al 1983). In this device, free carrier absorption in heavily doped quantum wells excited electrons over the barriers, with the cutoff energy being determined by the barrier height.

A superlattice avalanche photodetector (Figure 4) is another optical detector based on a multilayer structure (Capasso et al 1982). Although avalanche photodiodes generally have high gain and sensitivity, they are subject to avalanche multiplication noise. This problem is severe for GaAs diodes because the ionization rates for electrons and holes in the avalanche process are nearly equal, leading to more fluctuation in the gain. In the GaAs/(Al,Ga)As superlattice, on the other hand, the electron and hole avalanche ionization processes under strong field across the layers are modified in traversing barrier layers. This enhances the electron ionization rate over the hole rate and reduces multiplication noise.

### 4.1 Graded Barriers and Wells

The structures discussed above have involved layers with essentially square compositional and potential profiles. On the other hand, intentional grading of composition can generate graded potential profiles. An example of the use of intentional grading in very thin semiconductor layers is the unipolar rectifier which is produced by a sawtooth shaped barrier compositional profile (Gossard et al 1982) (Figure 5). Under applied voltage across the asymmetric barrier, rectification is produced. Symmetric barriers in the shape of triangles also can be produced and yield non-linear but symmetric current response to applied voltages.

Another application of compositional grading of thin semiconductor layers is graded base layers in heterojunction bipolar transistors (D L Miller et al 1983; Hayes et al 1983) (Figure 6). The speed of a bipolar transistor is enhanced by decreasing the time for a carrier to transit the base and by decreasing the base layer electrical resistance. The first of these desiderata suggests the need for a thin base layer and the second suggests the opposite. The idea of a graded base layer was introduced 25 years ago by

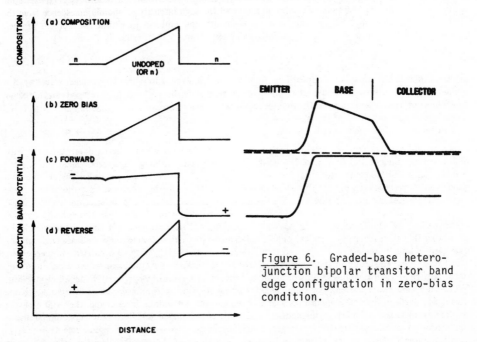

Figure 6. Graded-base hetero-junction bipolar transitor band edge configuration in zero-bias condition.

Figure 5. Sawtooth composition and potential profiles of unipolar rectifying heterostructure.

Kroemer (1957) and relieves the contradictory requirements by enhancing the electron speed across the base layer by means of the built-in quasi-electric field resulting from the graded composition and conduction band profile (Figure 6) (Levine et al 1983). Twenty-five years were necessary for materials to advance to the point where the idea could be implemented.

In addition to grading by changes in beam flux with molecular beam oven temperature or orifice size, there exists another means to grade potential profiles. This is growth with pulsed molecular beams, which can be turned on and off in less than a monolayer deposition time. This technique was used by Kawabe et al (1982) to generate (Al,Ga)As mixed compounds, quantum-wells with two Al levels and sinusoidal and triangular compositional profiles.

In fact, it is not necessary that the pulsing period be less than the monolayer growth time in order to achieve a desired effective band gap or effective band-edge potential. Electrons or holes may tunnel through several monolayers of barrier material and thus sense an average composition over a few monolayers. Layer growth with layer thicknesses of a few monolayers can produce bandgaps near the values of randomly mixed compounds. Aside from technical simplification by the longer shutter periods, these slightly thicker superlattices afford the cleaning and smoothing advantages of quantum wells referred to earlier.

R C Miller et al (1983) have used the pulsed beam technique with GaAs/(Al,Ga)As periods of $\approx 20\text{Å}$ to grow sequences of 300Å to 500Å-wide quantum wells with parabolic potential profiles. The average Al composition within the wells varies quadratically from zero at the well centers to $x_{Al}=0.30$ at the well edges and in the 200Å-wide (Al,Ga)As barriers. The average Al content is controlled by the Al shutter open time within each 20Å GaAs/(Al,Ga)As period. The expected quantum levels characteristic of a parabolically shaped quantum well profile are uniformly spaced in energy, proportional to $(n + 1/2)$, unlike the square well where the energy levels are quadratically spaced, proportional to $n^2$. The linear spacing is accurately realized in the parabolic well spectrum measured by photoluminescence excitation spectroscopy. The sharpness of the excitation lines and the high intensity of photoluminescence from the wells show that the 20 wells in the samples are smooth and uniform to within two percent of their thickness and that the structure is comparably radiative to GaAs and other GaAs structures.

Potential barriers in thin-layer semiconductor structures may be produced not only by compositional variations as discussed above, but also by finely varied dopant profiles. An early example is the camel diode (Shannon 1979), which employed a thin ion-implanted dopant layer to produce a rectification characteristic. More freedom in the design and fabrication of barriers by dopant profile was introduced in the n-i-p-i-n barrier (Malik et al 1980) grown by MBE and in the n-i-p-i superlattice (Dohler and Ploog 1980) also grown by MBE. In the n-i-p-i-n diode, a p-doped layer, placed asymmetrically in an i region between n terminal layers, provides an asymmetric barrier with rectification and with barrier heights up to nearly the semiconductor band gap. Structures placed back-to-back offer the possibility of a hot electron transistor with ballistic transport across center base layer (Malik et al 1981). When the n-i-p-i-n diode is superposed on a heterojunction pair, it affords a majority-carrier phototransistor (Chen et al 1981). The nipi superlattice has a corrugated band structure in which electrons and holes in adjacent layers have sub-bandgap energy differences. n and p layers may be contacted separately by selective contacts so that carrier concentrations, band modulation, absorption and luminescence all become voltage tunable (Kunzel et al 1982).

## 5. Quantum Well Electrical Transport

Charge carriers may be introduced into quantum well structures by optical creation of electrons and holes or by chemical doping. In this section we consider the conductivity in the doped quantum wells and the two-dimensional conductivity effects found in these structures. The two dimensional conductivity in quantum wells is a result of confinement in quantum states with no degree of freedom perpendicular to the wells but with free motion along the wells. Evidence of the two-dimensionality was demonstrated in GaAs/(Al,Ga)As quantum well structures in high magnetic fields by the angular dependence of Shubnikov-deHaas resistance oscillations (Chang et al 1977).

Electrons in GaAs quantum wells with greatly enhanced carrier mobilities and scattering lifetimes can be produced by a technique of selectively doping barrier layers in multiple quantum well structures (Dingle et al 1978). This technique, referred to as modulation doping, confines dopant atoms to barrier

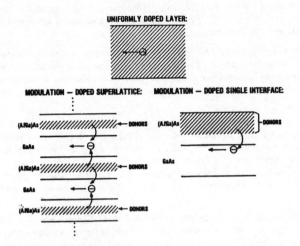

**Figure 7.** Dopant and carrier configuration of modulation-doped super-lattice and single interface and for uniformly doped material

layers by molecular beam dopant deposition only into barrier layers. Electrons from the dopants spontaneously transfer from the barrier layers to the adjacent well layers by virtue of the greater electron affinity and the consequent lower energy in the quantum wells. The spatial distribution of the dopant atoms and carriers is shown in Figure 7 and the relative positions of the band edges, dopant levels and electron levels for undoped, uniformly-doped and modulation-doped quantum wells are shown in Figure 8. Figure 7 also shows a configuration consisting of a single interface between a doped (Al,Ga)As barrier layer and an undoped GaAs layer in which carriers transferred from the (Al,Ga)As layer are bound in quantum states at the interface between the semiconductors. The quantum well in the single-interface structure is produced by the heterojunction potential step and the electric field of the ionized dopants. The potential configuration for this geometry is shown in Figure 9. The electron mobilities in modulation-doped structures are greatly enhanced relative to uniformly doped materials and with improved growth techniques and sample geometries, increasingly high mobilities have been observed (Figure 10). Low temperature mobilities in excess of $10^6 \text{cm}^2/\text{Vsec}$ have been measured (Hwang et al 1982). The source of the increased mobilities is the reduction in scattering from impurities in the current-carrying channels. The principal residual sources of scattering are phonons at the higher temperatures and ion scattering from the ionized dopants in the barriers at low temperatures. A calculation of the theoretical mobility for the simplified structure in which a sheet of ions of density $n_{ion}$ is separated by distance d from a sheet of mobile electrons of density $n_{el}$ is given by (Price 1982)

$$\mu_{ion} = \frac{16\sqrt{2\pi}\, e}{\hbar} \; \frac{n_{el}^{3/2}\, d^3}{n_{ion}}$$

The mobility increase with the spacer layer thickness d suggests the importance of inclusion of an undoped spacer layer in the barriers next to the conducting channel. The higher mobility with higher electron density $n_{el}$ is primarily a result of the greater Fermi wavevector at higher electron densities which results in smaller angular deviations in the elastic scattering process. The phonon scattering limit is being approached at temperatures from room temperature down to 1K, as shown in Figure 11, which compares recent mobilities (Hwang et al 1982) with mobility limits for the dominant phonon scattering processes. The increased mobility in the illuminated sample compared with the measurement in darkness is a result of a photoconductivity effect which increases the carrier concentration in the conducting channel by ionizing traps in the (Al,Ga)As barrier layer. The increase in mobility with concentration is also observed in gated structures in which carrier concentration is determined by the

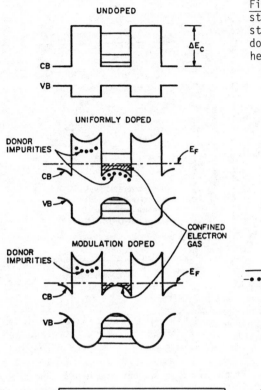

UNDOPED

$\Delta E_C$

CB

VB

UNIFORMLY DOPED

DONOR
IMPURITIES

$E_F$

CB

VB

MODULATION DOPED

DONOR
IMPURITIES

$E_F$

CB

VB

CONFINED
ELECTRON
GAS

Figure 8. Band edges, donor
states and conduction electron
states of undoped, uniformly-
doped and modulation-doped
heterostructure multilayers.

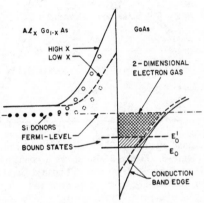

$A\ell_x\, Ga_{1-x}\, As$        GaAs

HIGH X
LOW X

2-DIMENSIONAL
ELECTRON GAS

Si DONORS
FERMI-LEVEL
BOUND STATES

$E_O^I$

$E_O$

CONDUCTION
BAND EDGE

Figure 9. Conduction band edge,
donor states and conduction
electron states at modulation-
doped single interface.

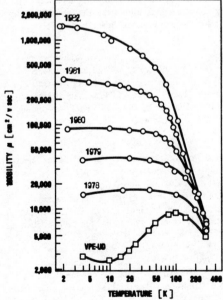

Figure 10. Highest reported
electron mobilities vs. year in
modulation-doped GaAs. Lowest
curve shows values for uniformly
doped GaAs. Structures vary but
all contain $\approx 10^{17}$ carriers/$cm^3$
in their conducting regions.

## SCATTERING MECHANISM

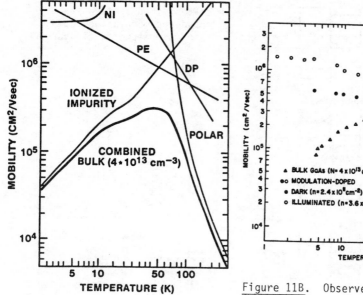

Figure 11A.   Mobility limits from dominant scattering processes.

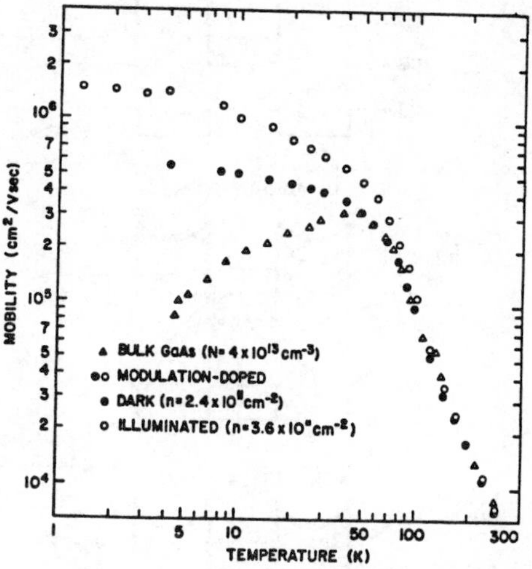

Figure 11B.   Observed mobilities from high-mobility modulation-doped single GaAs-(Al,Ga)As interface and from low- carrier-density GaAs.

voltage of a gate above or below the channel. When carrier concentrations increase to a point where a second band of quantum states, e.g. with two half wavelengths in the well, becomes occupied, the mobility is decreased somewhat by the new scattering channel which allows the electrons in the upper band to undergo large angle scattering by collision with electrons in the lower band (Störmer et al 1982).

A fundamentally different situation arises in heterostructures of GaSb and InAs, where both the conduction-band edge and valence-band edge of InAs lie below the valence-band edge of GaSb at the heterojunction interface. This produces electron states in the InAs and hole states in the GaSb at the heterojunction. In superlattices of the two materials, the bandgap can be less than the constituent bandgaps or even be negative, giving semimetallic behavior (Sakaki et al 1977). Both two-dimensional electron gases and two-dimensional hole gases can coexist at the interface.

### 5.1 Modulation Doped Devices

Modulation doped materials offer the potential for higher carrier velocities and lower electrical resistance than conventional structures and so are of interest in development of high speed electrical devices. Extensive effort is underway in the development of modulation doped field effect transistors and circuits, as discussed in the accompanying paper by Linh. High speed in such transistors requires fast acceleration and high velocity of electrons through channels under short gates. The entire mobility enhancement obtained at low fields in modulation-doped layers is not available for such applications, because these devices require high electric fields and high electron kinetic energy for fast operation. When electron energies exceed phonon energies in the modulation-doped structures, new scattering processes by phonon generation become possible, and reduction in the mobility enhancement occurs (Drummond et al 1982; Shah et al 1983) (Figure 12). Nonetheless, modulation doped FET's have shown substantially higher speed then uniformly doped structures. The highest speeds reported until now have been for modulation doped ring oscillator circuits with room-temperature propagation delays

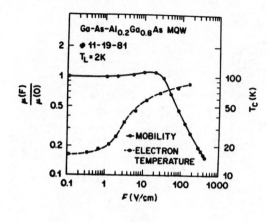

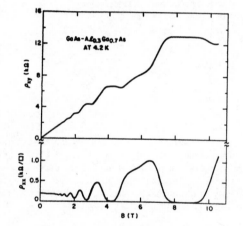

Figure 12. Field dependence of electron temperature and mobility of modulation-doped multiple quantum well structures.

Figure 13. Quantum Hall effect and resistance oscillations of modulation-doped GaAs-(Al,Ga)As single interface.

down to 12.2 ps and speed power products down to 13.6 fJ (Lee et al 1983). These are the highest speeds to date for a digital circuit and demonstrate that an advantage is provided by the modulation doped structures and their two-dimensional quantum-confined channels even without low temperatures and without the concomitant ultrahigh mobilities obtained at low temperature.

After demonstration of high speed in ring oscillators, it is important to develop more realistic logic circuits. Modulation-doped dividers based on flip-flops with fanouts up to three and with interconnects of practical lengths have now been fabricated and tested (Kiehl et al 1983; Nishiuchi et al 1983), and active development of more advanced devices is continuing.

### 5.2 Quantized Hall Effect

Although the low field Hall effect is routinely used to measure carrier concentrations and mobilities in uniform material and in quantum wells, remarkable new features develop in quantum well structures in the Hall response and conductivity at high magnetic fields. The quantum Hall effect was originally observed in silicon MOSFET's (von Klitzing et al 1980) and subsequently in modulation-doped GaAs/(Al,Ga)As heterostructure interfaces (Tsui and Gossard 1981). The effect is notable for the wide, flat plateaus in Hall resistance and the wide regions of nearly vanishing parallel resistance under variation of a magnetic field perpendicular to the two-dimensional electron gas at the heterostructure interface (Figure 13). The Hall resistance at the plateau is quantized in submultiples of $h/e^2$, where h is Planck's constant and e is the electron charge. The Hall resistance can be constant and reproducible to within a part in $10^8$ over a range of more than 1 kG of applied magnetic field (Tsui et al 1982). This

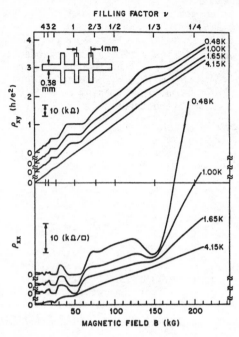

**Figure 14.** Extreme quantum limit Hall effect and resistance at high magnetic fields where lowest Landau level is fractionally filled to $\nu$ electrons per Landau level.

affords a new approach to measuring the fine structure constant, which depends on the exact value of $e^2/h$, and to providing an absolute standard of resistance. Recent interest has focussed on newly discovered features in the Hall and parallel resistance which develop at high fields under fractional occupation of the lowest cyclotron state (Störmer et al 1983) (Figure 14). Under these circumstances, the electrons order in a new state which is apparently an incompressible electron liquid whose excitations are fractionally charged quasi-electrons and quasi-holes (Laughlin 1983).

### 6. Other Areas of Research

The excitations which can be diffracted from superlattices are not restricted to electromagnetic waves. The possibility of phonon diffraction by multilayers was explored using $GaAs(Al_{0.5}Ga_{0.5}As)$ superlattices (Narayanamurti et al 1980). Phonons transmitted through the superlattice showed a substantial attenuation at the wavelength where the 50Å layers of the superlattice were quarter phonon wavelengths. Such a device forms a basis for coherent phonon optics and the mirrors for possible phonon lasers.

Another area of potential superlattice and quantum well research and application in which work has started is the development of one-dimensional quantum well structures in which particles are bound in two dimensions, leaving only one free dimension for unconfined motion. Several approaches may be made to one dimensional structures. One avenue which has shown progress is the production of one-dimensional channels by means of photolithographic etching of narrow ridges along the surface of a two-dimensional GaAs quantum layer structure followed by overgrowth with a cladding layer of (Al,Ga)As for low loss confinement at the edges of the ridges (Petroff et al 1982). In a multilayer stack this has produced 200Å layers in a triangular ridge in which the narrowest layer is ≈200Å wide. Another approach has been suggested in which an etched groove in a layered heterostructure reveals the edge of a quantum layer on which an insulated gate or doped semiconductor barrier could be deposited to produce a one-dimensionally quantized inversion layer (Sakaki 1980). Fabrication of such a structure has not yet been reported.

These new possibilities together with the ones covered in the earlier sections of the paper suggest that quantum wells and superlattices in semiconductors are greatly expanding the range of behavior and performance attainable from semiconducting materials.

I wish to acknowledge the collaboration of W. Wiegmann, H. L. Störmer, D. C. Tsui, R. A. Logan, R. Dingle, J. M. Worlock, A. Pinczuk, J. Shah, V. Narayanamurti, R. C. Miller, P. M. Petroff, W. T. Tsang, M. B. Panish, D. A. Kleinman, C. Weisbuch, H. M. Gibbs, S. L. McCall, T. N. C. Venkatesan, A. Passner, J. L. Jewell, F. Capasso, J. R. Hayes, R. J. Malik, C. L. Allyn, M. A. Chin, M. Paalanen, T. Haavasoja and G. E. Derkits in work described here.

## REFERENCES

Burnham R D, Lindstrom C, Paoli T L, Scifres D R, Streifer W and Holonyak Jr N 1983 Appl. Phys. Lett. *42* 937

Capasso F, Tsang W T, Hutchinson A L and Williams G F 1982 Appl. Phys. Lett. *40* 38

Chang L L, Esaki L and Tsu R 1974 Appl. Phys. Lett. *24* 593

Chang L L, Sakaki H, Chang C A and Esaki L 1977 Phys. Rev. Lett. *38* 1489

Chen C Y, Cho A Y, Garbinski P A, Bethea C G and Levine B F 1981 Appl. Phys. Lett. *39* 340

Chiu L C, Smith J S, Margalit S, Yariv A and Cho A 1983 Infrared Physics *23* 93

Coleman J J, Dapkus P D, Laidig W D, Vojak B A and Holonyak Jr N 1981 Appl. Phys. Lett. *38* 63

Dingle R 1975 Festkorperprobleme ed H J Queisser (Oxford: Pergamon) Vol XV, pp 21-48

Dingle R, Gossard A C and Wiegmann W 1975 Phys. Rev. Lett. *34* 1327

Dingle R and Henry C H 1976 U.S. Patent 3,982,207 9/1976

Dingle R, Störmer H L, Gossard A C and Wiegmann W 1978 Appl. Phys. Lett. *33* 665

Dohler G H and Ploog K 1979 Progr. Crystal Growth Charact *2* 145

Drummond, T J, Kopp W, Morkoc H and Keever, M 1982 Appl. Phys. Lett. *41* 277

Dupuis R D, Hartman R L and Nash F R 1983 41st Annual Device Research Conference (Burlington) paper IVB-2

Esaki L and Tsu R 1970 IBM J. Rev. Dev. *14* 61

Gibbs H M, McCall S L, Venkatesan T N C, Gossard A C, Passner A and Wiegman W 1979 Appl. Phys. Lett. *35* 451

Gibbs H M, Tarng S S, Jewell J L, Weinberger D A, Tai K, Gossard A C, McCall S L, Passner A and Wiegmann W 1982 Appl. Phys. Lett. *41* 221

Gossard A C, Petroff P M, Wiegmann W, Dingle R and Savage A 1976 Appl. Phys. Lett. *29* 323

Gossard A C, Brown W, Allyn C L and Wiegmann W 1982 J. Vac. Sci. Tech. *20* 694

Hayashi I, Panish M B and Foy, P W 1969 IEEE J. Quantum Electron. *QE-5* 211

Hayes J R, Capasso F, Gossard A C, Malik R J and Wiegmann W 1983 Electron. Lett. *19* 410

Holonyak Jr N, Laidig W D, Vojak B A, Hess K, Coleman J J, Dapkus P D and Bardeen J 1980 Phys. Rev. Lett. *45* 1703

Hwang J C M, Kastalsky A, Störmer H L and Keramidas VG 1982 MBE-CST-2 (Tokyo) postdeadline paper

Kawabe M, Matsuura N and Inuzuka H, 1982 Jap. Jour. Appl. Phys. *21*, L447

Kiehl R A, Feuer M D and Hendel R H 1983 41st Annual Device Research Conference (Burlington) paper IVA-3

Kroemer H 1957 RCA Rev. *18* 332

Kroemer H 1963 Proc. IEEE *51* 1782

Kunzel H, Dohler G H, Ruden P and Ploog K 1982 Appl. Phys. Lett. *41* 852

Laughlin R B 1983 Phys. Rev. Lett. *50* 1395

Lee C P, Miller D L, Hou D and Anderson R J 1983 41st Annual Device Research Conf. (Burlington) paper IIA-7

Levine B F, Bethea C G, Tsang W T, Capasso F, Thornber K K, Fulton R C and Kleinman D A 1983 Appl. Phys. Lett. *42* 769

Ludowise M J, Dietze W T, Lewis C R, Camras M D, Holonyak Jr N, Fuller B K and Nixon M A 1983 Appl. Phys. Lett. *42* 487

Malik R J, AuCoin T R, Ross R L, Board K, Wood C E C and Eastman L F 1980 Electron Lett. *16* 836

Malik R J, Hollis M A, Eastman L F, Woodard D W, Wood C E C and AuCoin T R 1981 Eighth Biennial Conf. on Active Microwave Semiconductor Devices and Circuits

Miller D A B, Chemla D S, Eilenberger D J, Smith P W, Gossard A C and Tsang W T 1982 Appl. Phys. Lett. *41* 679

Miller D L, Asbeck P M, Anderson R J and Eisen F H 1983 41st Annual Device Research Conf. (Burlington) paper IIA-1

Miller R C, Kleinman D A and Gossard A C Inst. Phys. Conf. Ser. No. 43 1979 Chap. 27, pp. 1043-1046

Miller R C, Kleinman D A, Nordland Jr. W A and Gossard A C 1980 Phys. Rev. B *22* 863

Miller R C, Kleinman D A, Tsang W T and Gossard A C 1981 Phys. Rev. B *24* 1134

Miller R C, Gossard A C and Wiegmann W 1983 (to be published)

Narayanamurti V, Störmer H L, Chin M A, Gossard A C and Wiegmann W 1980 Phys. Rev. Lett. *43* 1536

Nishiuchi K, Mimura T, Kuroda S, Hiyamizu S, Nishi H and Abe M 1983 41st Annual Device Research Conf. (Burlington) paper IIA-8

Osbourn G C, Biefeld R M and Gourley P L 1982 Appl. Phys. Lett. *41* 699

Petroff P M, Weisbuch C, Dingle R, Gossard A C and Wiegmann W 1981 Appl. Phys. Lett. *38* 965

Petroff P M, Gossard A C, Logan R A and Wiegmann W 1982 Appl. Phys. Lett. 41 635

Pinczuk A and Worlock J M 1982 Surface Sci. *113* 69

Price P J 1982 Surf. Sci. *113* 199

Rezek E A, Chin R, Holonyak Jr. N, Kirchoefer S W and Kolbas R M 1980 Jour. of Electronic Materials *9* 1

Sakaki H, Chang L L, Ludeke R, Chang C A, Sai-Halasz G and Esaki L 1977 Appl. Phys. Lett. *31* 211

Sakaki H 1980 Jpn. Jour. Appl. Phys. *19* L735

Schulman J N and McGill T C 1981 Phys. Rev. B *23* 4149

Shah J, Pinczuk A, Störmer H, Gossard A C and Wiegmann W 1983 Appl. Phys. Lett. *42* 55

Shannon J M 1979 Appl. Phys. Lett. *35* 63

Störmer H L, Gossard A C and Wiegmann W 1982 Solid State Comm. *41* 707

Störmer H L, Chang A, Tsui D C, Hwang J C M, Gossard A C and Wiegmann W 1983 Phys. Rev. Lett. *50* 1953

Temkin H, Alavi K, Wagner W R, Pearsall T P and Cho A Y 1983 Appl. Phys. Lett. *42* 845

Tsang W T 1981 Appl. Phys. Lett. *39* 786

Tsang W T 1982 Collected Papers of MBE-CST-2 (Tokyo: Japan Society of Applied Physics pp. 75-79

Tsui D C and Gossard A C 1981 Appl. Phys. Lett. *38* 550

Tsui D C, Gossard A C, Field B F, Cage M E and Dziuba R F 1982 Phys. Rev. Lett. *48* 3

Von Klitzing K, Dorda G and Pepper M 1980 Phys. Rev. Lett. *45* 494

Weisbuch C, Miller R C, Dingle R, Gossard A C and Wiegmann W 1981 Solid State Comm. *37* 219

*Inst. Phys. Conf. Ser. No. 69*
*Paper presented at ESSDERC/SSSDT 1983, Canterbury 13–16 Sept. 1983*

# Superlattices and electron devices

NUYEN T. LINH
THOMSON-CSF Central Research Laboratory
91401 ORSAY FRANCE

Abstract. The first work on periodic semiconductor superlattice based on the idea of subdivision of Brillouin zones into series of mini-zones has not led to practical applications. But the realization of superlattice by molecular beam epitaxy has revealed to physicists powerful technologies by which new classes of semiconductor can be tailored. Electron devices made with these thin films have shown excellent performances. This review describes these devices and shows how their performances can be correlated with the physical properties of these new classes of semiconductor. Conclusion is given on the tight cooperation between physicists and engineers to build up new ideas and new devices.

## 1. Introduction

Electron devices with higher and higher performances are forseen for the application of future electronic systems such as optical fiber communication, satellite transmission, radar links, direct satellite broadcasting, super-computer, high speed signal processor... (Fig. 1). For example, high power emitters and low noise amplifiers are necessary for satellite communication and radar links, very high speed transistors constitute the basic device for super-computers and high speed signal processors, while optical fiber communication is dependent on the quality of semiconductor lasers and detectors.

Novel devices which have recently been created for these application are based on the utilization of thin hetero-layers whose basic structure is more or less derived from the concept of superlattice. What is a superlattice from its original meaning?

Superlattices have been conceived as man-made one-dimensional *periodic* structures constituted by different semiconductors for which the rather long period ($\sim$ 100 Å) would produce the subdivision of Brillouin zones into series of mini-zones (Esaki and Tsu 1969). Fig. 2 shows the schematic band diagram and band structure of GaAs-Al$_x$Ga$_{1-x}$As superlattice. Because of the existence of these mini-zones, interesting effects were predicted in these new semiconductor "crystals", in particular differential negative resistance and Bloch oscillations. In fact such properties have not been experimentally demonstrated. But practical realizations of superlattices by molecular beam epitaxy have revealed to physicists powerful technologies by which new classes of semiconductor based on thin films, can be tailored. Multiple quantum wells (MQW) (Dingle et al 1974) modulation-doped heterojunctions (Dingle et al 1978), superlattice

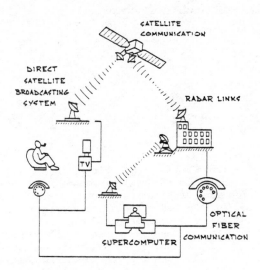

Fig. 1   Future electronic systems need
high performance electron devices

photo avalanche structures (Capasso et al 1982), nipi superlattices
(Döhler 1972)... are some of these multilayer structures. Electron devices
(laser diodes, field effect transistor, photodetector)  made with these
structures have shown excellent performances. The objective of this
review is to describes these devices and to show how they are related to
the concept of periodic superlattice.


2. Multiple quantum wells and laser diodes

Instead of considering a superlattice as a periodic structure
giving rise to interesting transport properties, Dingle et al (1974)
treated the thin multilayer structures as a multiple potential well.
Fig. 3 shows the wells formed by GaAs layers whereas barriers are due to
$Al_xGa_{1-x}As$, the larger band gap material. If the width $L_z$ of the potential
well is short as compared to the electron (or hole) de Broglie wavelength,
the energies of electron  (or hole) are quantized in two-dimensional
bands whose energy levels are given by

$$E = E_{//} + E_n \qquad E_n = \frac{h^2}{2m}\left(\frac{n\pi}{L_z}\right)^2 \qquad n = 1, 2, 3 \cdots$$

where $E_{//}$ is the energy associated with motion in the x-y plane, $E_n$ is the
energy associated with quantization in the z direction, of the $n^{th}$ level, h
the Plank constant, m the effective mass of electron (or hole).

The two-dimensional character of the well also modifies the density of states which changes from a parabolic variation to a stepped
variation as represented in Fig. 3.

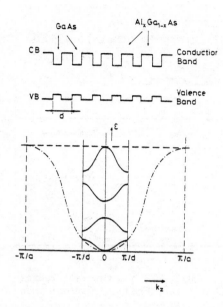

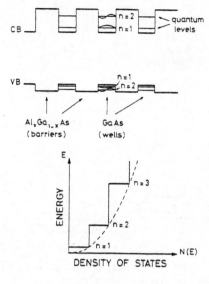

Fig. 2  Schematic band diagram
and band structure of a GaAs-
Al$_x$Ga$_{1-x}$As periodic superlattice
(after Esaki et al 1969)

Fig. 3  GaAs-Al$_x$Ga$_{1-x}$As multiple
quantum well structure. The step-
like density of states is characteristic of a two-dimensional
system.

The two properties – quantized subbands and step-like density of
states - lead to important applications in laser diodes. As a matter of
fact, the quantization of energy levels in quantum wells (QW) gives rise
to transitions between subbands with the selection rule $\Delta n = 0$, as noted
by Dingle (1975). These transitions have been experimentally observed
for the first time by absorption measurements (Fig. 4  ). Referring to
this figure one can immediatly deduce that the transition energies in QW
are higher than in bulk GaAs; therefore the wavelength associated with
these transitions are shorter.

Application in the domain of visible lasers is evident. By using
GaAs- Al$_x$Ga$_{1-x}$As MQW as active layer of double heterostructure lasers
Camras et al (1982) have fabricated visible lasers with a wavelength as
short as 6500 Å. Conventional lasers with AlGaAs as active layer can work
down to 7150 Å but their characteristics (CW operation, output power...)
are poorer than those of MQW lasers. It was believed that part of the
improved performance could be attributed to quantum size effect.

However, the interest of QW laser is certainly not limited to the
realization of short wavelength laser. The step-like density of states of
a QW can also be used to improve the threshold current of a laser.

The first experimental evidence of the threshold current $J_{th}$ reduction in QW laser was observed by Tsang (1981). Fig. 5 represents the

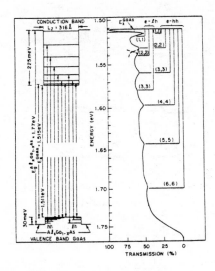

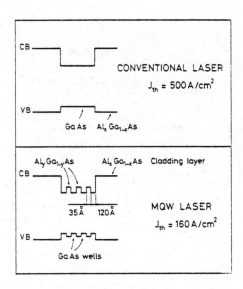

Fig. 4  Transitions in QW well
and comparison with experimental
transmission spectrum (after
Dingle 1975)

Fig. 5  Multiple quantum well lasers
present lower threshold current than
conventional one.

schematic band diagram of the MQW laser realized by Tsang and compares
it to that of a conventional laser. The improvement in $J_{th}$ (250 A/cm$^2$ with
respect to 500 A/cm$^2$ for the best conventional laser), which results
from many efforts of research on superlattice (since 1969) and MQW (since
1974) is a success of a long term research programme. Better results have
now been obtained with single QW lasers in which the well is positionned
in a graded-refrective index (GRIN) separate-confinement-heterostructure
(SCH) (Tsang et al 1982) (Hersee et al 1982). GRIN-SCH-QW lasers offer a
threshold current density as low as 121 A/cm$^2$ or a threshold current of
2.5 mA. The latter value is 3 times lower than in conventional lasers and
is the lowest ever reported.

     The two-dimensional character of the QW also improves the tempera-
ture sensitivity of the laser (Hess et al 1980 and Arakawa et al 1982).

## 3. Modulation-doped superlattices and field effect transistors

     Modulation-doped superlattices are multilayer structures (for
example GaAs-Al$_x$Ga$_{1-x}$As) in which the smaller band gap semiconductor
layers are undoped and the larger band gap ones are n-doped (Dingle et al
1978). If the electron affinity of the smaller band gap material is higher
than that of the larger band gap material (which is the case for GaAs and
Al$_x$Ga$_{1-x}$As) electrons are shifted towards the smaller gap semiconductor
leaving the parent donor impurities in the larger gap semiconductor.
Electrons are threfore spatially separated from ionized impurities. Fig. 6
represents the schematic band diagram of a modulation-doped superlattice.
Because of the charge separation, some band bending is observed both sides
of the heterojunction.

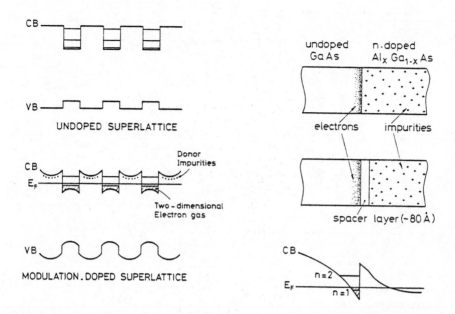

**Fig. 6** Band structure of modulation superlattice (after
Dingle et al 1978)

**Fig. 7** Selectively doped
single heterojunction.

   The electron–impurity spatial separation is a new concept in
semiconductors, since up to now the doping of semiconductors supplies both
free electrons and ionized impurities in the same material. The spatial
separation of electrons and impurities reduces their coulombian scattering and therefore enhances the electron mobility, particularly at low
temperature where phonon scattering is reduced. Dingle et al (1978) observed for the first time, mobilities of 6000 and 20 000 $cm^2v^{-1}s^{-1}$respectively
at 300 and 4 K, for modulation–doped superlattices doped to the range of
$10^{17}$-$10^{18}cm^{-3}$. Notice that the word "superlattice" is used here while no
periodicity is required.

In fact, the notion of electron–impurity spatial separation is not limited
to multilayers. Selectively-doped single heterojunctions in which GaAs is
undoped and $Al_xGa_{1-x}As$ is n-doped also present the reduced coulombian
scattering and therefore an enhancement in electron mobility. To increase
the spatial separation of  charges, an undoped $Al_xGa_{1-x}As$ spacer layer
can be grown between the undoped GaAs and the n-doped $Al_xGa_{1-x}As$ layers.
The thickness of the spacer layer can vary between 20 and 200 Å. Fig. 7
schematically represents a selectively-doped single heterojunction with
a spacer layer and the shape of the conduction band. The interface QW
can be described as a triangular potential well (Delagebeaudeuf et al
1982a) therefore the longitudinal quantized energies are well approximated by

$$E_n = \left(\frac{\hbar}{2m}\right)^{1/2} \left(\frac{3}{2}\pi q \mathcal{E}\right)^{2/3} \left(n + \frac{3}{4}\right)^{2/3} \qquad n = 0, 1, 2 \cdots$$

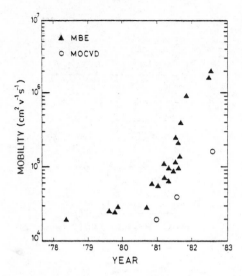

Fig. 8  The improvement of low
temperature 2 DEG mobility is
impressive

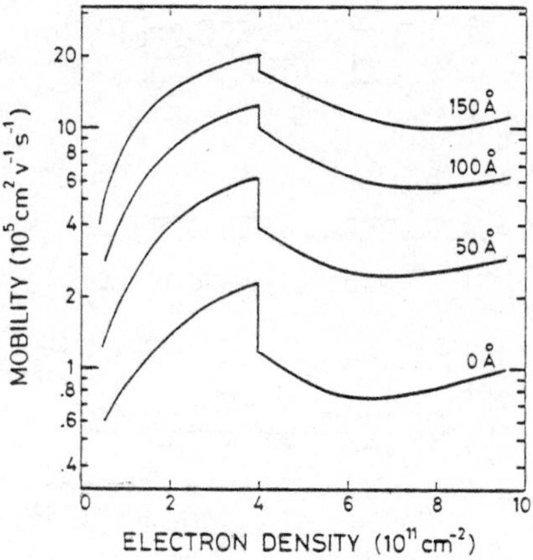

Fig. 9  Theoretical results on
mobilities in 2 DEG (after
Fishman et al 1983)

where $\mathcal{E}$ is the electric field at the interface. Because of the high electron mobility and the quantization effect in the direction perpendicular
to the interface, electrons which are located in the GaAs QW are said to
form a two-dimensional electron gas (2DEG).

By improving the material quality and using a thick spacer layer,
several workers have obtained mobilities as high as $10^6-2\times10^6 cm^2 v^{-1}s^{-1}$ at
4 K (Hiyamizu 1982 and Hwang et al 1982). Fig. 8 represents the evolution
of the low temperature mobility obtained in different laboratories on
modulations doped superlattices or heterojunctions since 1978 : the
impressive progress noted on MBE layers is attributed to the improvement
of the background  impurity concentration. The mobility in MOCVD layers
whose best result is 162 000 $cm^2v^{-1}s^{-1}$ (Hersee et al 1982b) is certainly
limited by background impurities in the AlGaAs layers, as it has been
theoretically shown by Bastard et al (1982). For example these latter
authors have found that      a background impurity concentration of $\sim 10^{14}$
$cm^{-3}$ would give a mobility of $3\times10^6 cm^2v^{-1}s^{-1}$ which will be reduced to
$2\times10^5 cm^2v^{-1}s^{-1}$ if the impurity concentration in the spacer is increased
to  $5\times10^{16} cm^{-3}$.

Extremely high mobility has also been theoretically predicted by
Fishman et al (1983). Fig. 9 shows the influence  of electron density in
the 2 DEG and the spacer thickness on the low temperature mobility. The
drastic drop of mobility at electron density of $4\times10^{11} cm^{-2}$ is due to
intersubband scattering first predicted by Mori et al (1980).

The reduction of coulombian scattering in the modulation-doped 2DEG
which gives rise to extremely high mobility, has promoted fantastic
studies of quantum Hall effect at high magnetic field (Tsui et al 1982a
and 1982b). This interesting subject will not be discussed in this paper

which deals with electron devices.

High electron mobility has led to the idea that the transistor working with this 2DEG would present high speed and high cut-off frequency. This transistor labelled high electron mobility transistor (HEMT) (Mimura et al 1980) or two-dimensional electron gas field effect transistor (TEGFET) (Delagebeaudeuf et al 1980) has been simultaneously developed by Fujitsu and Thomson-CSF Laboratories. The cross-sectional view of a conventional GaAs FET and a TEGFET is represented in Fig. 10. Both of them use a Schottky gate to control the drain current. Fig. 11 compares the mobility of a TEGFET to that of a conventional GaAs FET structures. It can be noted that the mobility enhancement reaches a factor of 2 at 300K and 15 to 20 at 77K.

But it cannot be concluded through this mobility enhancement that the TEGFET is "twice better" and "20 times better" than the conventional

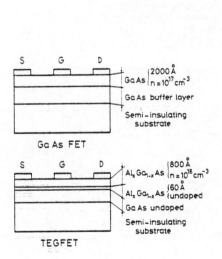

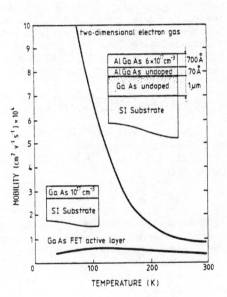

Fig. 10 Conventional GaAs FET and TEGFET structures

Fig. 11 TEGFET exhibits extremely high electron mobility compared to conventional GaAs FET

GaAs FET, because in a short gate length FET, the performance of the transistor (noise, cut-off frequency...) is directly related to the electron velocity but not the electron mobility.. According to Delagebeaudeuf et al (1982a and 1982b) the drain current of a TEGFET varies linearly with the gate voltage :

$$I_{DS} \sim g_m (V_G - V_T)$$

$$g_m = \frac{g_{m_o}}{1 + R_s g_{m_o}}$$

$$g_{m_o} = \frac{\varepsilon_2 \, v_s \, Z}{d_2}$$

where $I_{DS}$ is the drain current $V_G$ the gate voltage, $V_T$ the pinch-off voltage, $g_m$ the transconductance, $g_{mo}$ the intrinsic transconductance, $R_s$ the source resistance, $\varepsilon_2$ the $Al_xGa_{1-x}As$ permittivity, $d_2$ the effective thickness of the $Al_xGa_{1-x}As$ layer, Z the gate width and $v_s$ the peak electron velocity in the 2DEG.

DRUMMOND et al (1982) have estimated by Monte Carlo calculations that the electron velocity in the 2DEG is higher than in bulk n-doped GaAs by a factor of 60% at 300K and 200% at 77 K. In fact, this consideration of velocity is made in its steady-state regime. It is now well known that in short gate length FET ($\lesssim 1\mu m$) the non-stationary (or dynamic) regime has to be taken into account. The non-stationary regime gives rise to overshoot effect, i.e., electron velocity is much higher than in the steady-state regime. This overshoot phenomenon is the 2DEG has been studied by CAPPY et al (1982) and Murdares et al (1983), who have shown that overshoot effect is more important in TEGFET than in GaAs FET.

In addition to the electron velocity enhancement, the other advantages which can be a priori predicted for the TEGFET are :

1 - electrons in the 2DEG are located in a deep potential well, and therefore cannot easily be injected out of the well unlike the case of a GaAs FET. Low output conductance observed on TEGFETs confirm this prediction.

2 - In a TEGFET, where electrons are moving in a high purity material, many electrons are then surrounded by few impurities. Hence screening effect contributes to reduce the interaction between electrons and impurities and therefore enhances mobility and velocity. In other term, for a given impurity concentration, the mobility (and velocity) depends on the electron concentration. In the 2DEG system, this property can be verified by measuring the variation of mobility of a TEGFET versus the gate voltage i.e. the electron concentration (Wallis 1980). By assuming that the electron concentration dependence of the mobility obeys the equation :

$$\mu = \mu_o (n_s)^k$$

Delagebeaudeuf et al (1982b) have found that the best fit with experimental data gives $k \sim 1.5$ (Fig. 12). Screening effect perfectly explains why the low temperature mobility of a modulation doped structures exceeds that of a extremely high purity GaAs material (linh 1983).

Reduced coulombian scattering, screening effect and deep potential well are favourable to transport properties in a FET for high frequency, high speed and low noise operation.

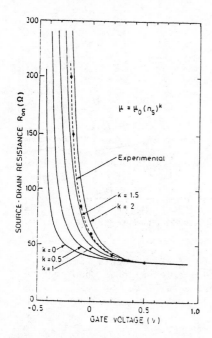

**Fig. 12** The 2DEG mobility is a function of sheet electron concentration : screening effect plays an important role in mobility

Low noise TEGFETs were first fabricated by Laviron et al. (1981) . Both E-mode and D-mode transistors were obtained  and studied. With a gate length of 0.8 μm and a FET geometry without recessed gate, microwave performances were promising. At 10 GHz, the noise figure was 2.3 dB and the associated gains were 10.3 and 7 dB for D-mode and E-mode transistors respectively.

Niori et al (1983) working with recessed gate transistors have achieved better performances with a 0.5 μm gate length and 2 μm source-to-drain spacing. Their transistors present at 8, 11.3 and 20 GHz, noise figures of 1.3, 1.7 and 3.1 dB respectively and associated gains of 13, 11.2 and 7.5 dB.

A four stage amplifier using the HEMT has given, at 20 GHz, a noise figure of 3.9 dB with a power gain of 30 dB.

More recently Fujitsu's group reported a noise figure of 1.4 dB at 12 GHz for a 0.5 μm gate length device Joshin et al (1963). This represents a good result compared to conventional GaAs FETs.

Our latest results on 0.55 μm gate length TEGFETs are (Fig. 13)

$$
\begin{array}{ll}
\text{10GHz} \\
\text{(300K)}
\end{array}
\left\{
\begin{array}{lll}
\text{NF} & = & 1.26\text{dB} \\
G_{ass} & = & 12 \ \text{dB} \\
G_{max} & = & 14.5 \ \text{dB}
\end{array}
\right.
\qquad
\begin{array}{ll}
\text{10GHz} \\
\text{(120K)}
\end{array}
\left\{
\begin{array}{lll}
\text{NF} & = & 0.25\text{dB} \\
G_{ass} & = & 15 \ \text{dB} \\
G_{max} & = & 24 \ \text{dB}
\end{array}
\right.
$$

$$
\text{17.5GHz}
\left\{
\begin{array}{lll}
\text{NF} & = & 2.3\text{dB} \\
G_{ass} & = & 7.1\text{dB} \\
G_{max} & = & 9 \ \text{dB}
\end{array}
\right.
$$

These performances, particularly at 17.5 GHz, are better than that of conventional GaAs MESFETs. Moreover it has to be noticed that the TEGFET is not optimized yet.

To estimate the optimum performance of the TEGFET we can used the semi-empirical Fukui equation which gives the minimum noise figure dependence on parasitic elements :

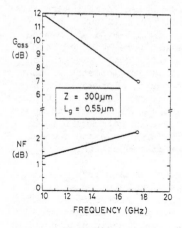

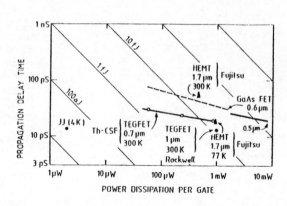

Fig 13   Low noise TEGFETs already present excellent performance while its technology is not optimized yet.

Fig 14 TEGFET presents higher speed and lower power than conventional GaAs FET.

$$F_{min} = 1 + 2\pi C_{gs} f K_F \sqrt{\frac{R_s + R_g}{g_{mo}}}$$

where $C_{gs}$ is the gate to source capacitance, f the frequency and $K_F$ is a fitting factor which is related to the material quality. In conventional GaAs FET, $K_F \geqslant 2.5$. Our determination of KF on TEGFET, and those performed in Fujitsu's Group, have shown that $K_F$ (TEGFET) $<$ 2 and can be as low as 1.5. While this result is not completely understood yet, it seems that the reduced dimensionality of the electron population in the TEGFET could be responsible for the low value of the $K_F$ factor (Linh 1983b and Linh et al 1983).

The first TEGFET integrated circuit fabricated by Mimura et al (1981) exhibited a propagation delay time of 56.5 ps at 300 K and 17.1 ps at 77 K. The improvement by a factor of 3.3 between room temperature and 77K is the consequence of the enhancement of mobility and velocity of the 2DEG at low temperature. By using a gate length of 0.7 $\mu$m Tung et al (1982) observed at room temperature a propagation delay time as low as 18.4 ps and a power dissipation of 0.9 mW. The same circuit can operate at 32.ps with 62 $\mu$W. Recently Lee et al (1983) have obtained a delay time of 12.2 ps at 300K with a power dissipation of 1.1 mW (gate length ~ 1 $\mu$m) and Tung et al (1983) have reached very low power dissipation operation (41 $\mu$W with delay time of 34 ps).

These values represent the best figures of merit reported so far on integrated circuits working at 300 K. They are approching those of Josephson junction devices. In Fig. 14 are reported the power-delay characteristics of the highest performance devices. It can be noted that TEGFETs are not only faster than conventional GaAs FETs, but they also present a smaller power dissipation. This latter property is an important parameter in very large scale integration where power dissipation can represent a limiting factor. The low power dissipation is also a fundamental parameter for electronic systems working in space.

More complicated and realistic circuits have recently been realized: frequency dividers. With a gate length of ~ 1 $\mu$m TEGFET frequency dividers already operate up at 3.2 GHz with D-type flip-flop (Kiehl et al 1983) and at 5.5 GHz in master-slave configuration (Nishiuchi et al 1983). These figures are better than those obtained on conventional GaAs FET, and are comparable to GaAs FET circuits having gate length twice shorter. Both results observed on ring oscillators and frequency dividers confirm that the electron velocity is higher in the TEGFET than in conventional FET device.

## 4. Superlattice and avalanche photodiodes

A large difference between the ionization coefficient for electron and holes is an essential requirement for low noise avalanche photodiode (APD). It is well known that in bulk GaAs, the ionization coefficient for electrons and holes are approximately equal. The utilization of a GaAs-$Al_xGa_{1-x}$As multilayer can change the effective impact ionization rates for electrons and holes (Capasso et al 1982). Fig. 15 shows the band diagram of a GaAs-$Al_xGa_{1-x}$As superlattice APD under bias. Whenever an electron crosses an $Al_xGa_{1-x}$As-GaAs interface it gains an energy $\Delta E_C$ corresponding to the conduction band discontinuity. At the interface holes have their energy increased by $\Delta E_V$ . Because of the large difference between $\Delta E_C$ and $\Delta E_V$ ($\Delta E_C/\Delta E_V = 4$) electrons gain more energy than holes. Then their impact ionization rate is becoming higher. Experiments confirm this consideration and a ratio of impact ionization rate $\alpha/\beta$ of 10 has been measured on superlattice APD.

Similar effect can be obtained with a staircase structure in which graded multilayer structures, are used.

## 5. Other devices

The three examples cited above correspond to devices which have clearly demonstrated their high performances. Many other devices based on the utilization of multilayer structures present interesting potential of application. Among these, we can cite :

- nipi superlattice device : nipi superlattice first proposed by Döhler (1972, 1979), are alternate layers with different doping types, n and p, in which novel properties such as tunability of the carrier concentration, of the band gap as well as of the subband separation and the absorption soefficient are predicted.

-planar doping device : Malik et al (1980) have proposed to use as a rectifier an npn junction in which the p layer is as thin as few tens of Å. This type of planar doping diode has been found to be an ultra-high speed photodiode with rise time of less than 30 ps (Chen et al 1981).

- real space transfer device : the Gunn diode is based on the transfer of electrons from the central valley to the upper valley (k-space transfer mechanism). In a GaAs/AlGaAs multilayer structure hot electron moving in the plane parallel to the heterojunction interface, can be transferred from the GaAs wells to the AlGaAs layers where their mobility is lower. Negative differential resistance is then predicted (Hess et al 1979), with transfer speed of the order of 10 ps. Some experimental data have confirm this prediction (Keever et al 1981).

- Strained superlattice devices : Up to now, all heterojunction devices are made of materials having the same lattice parameter, for example GaAs/AlGaAs, $Ga_{0.47}In_{0.53}As/InP$ etc... But it is known that is quite possible to epitaxially grow a material whose lattice parameter is different from that of the substrate. If the epitaxial layer is thin enough so that strains can be accomodated in the layer, no misfit dislocation is created at the interface between substrate and layer. With a lattice mismatch $\Delta a/a$ of $10^{-2}$ , the maximum thickness of the undislocated layer is about 100 Å. Such a strained layer exhibits modified band gap energy (Osbourn 1982), and electron effective mass. Applications for opto-electronic devices and high-speed transistors are foreseen.

## 6. Conclusion

Strictly speaking, and  considering that a superlattice is a one-dimensional *periodic* structure, there is no superlattice device. But extending the term  superlattice to *non-periodic* heterolayers and to superthin ($\sim$10 - 100 Å) multilayer structures one can conclude that the superlattice has  led to the fabrication of very high performance electron devices. QW lasers, superlattice APD, planar doped photodiode will be used in optical fiber communication, TEGEFT in radar links, satellite communication and supercomputer. Fig. 16 summarizes the superlattice story.

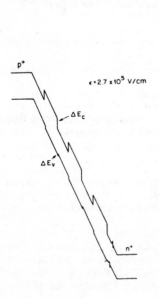

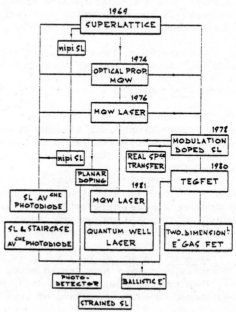

Fig 15 Energy band diagram of a superelattice APD (after Capasso et al 1982)

Fig. 16  Superlattice story

Approximately 10 years separate the first publication on superlattice and the realization of high performance MQW laser and TEGFET. During these ten years, ideas were moving from one concept to another but with a common objective : how to build new classes of semiconductor by using powerful modern technologies. Technologies, mainly molecular beam epitaxy in this superlattice story, have played a key role in the progress, not only in the fabrication of devices but also in the discovery of new physical phenomena. The fractional quantum Hall effect is one example.

The success of superlattice studies illustrates   efficient cooperation between physicists   and engineers for the benifit of both of them.

## Acknowledgements

The author would like to thank the DRET for     support on the TEGFET studies. Stimulating discussions with B. de Cremoux on QW lasers are appreciated. The author gratefully thanks R. Dingle and F. Capasso for permission to reproduce figures from their publications, and his co-workers, especially D. Delagebeaudeuf , P. Delescluse, M. Laviron, P.N. Tung, J. Chevrier, for their contribution.

## References

Arakawa Y and Sakaki H 1982 Appl. Phys. Lett. 40 939

Bastard G and Mendez EE 1982 Private Communication
Camras MD, Holonyak N, Hess K, Coleman JJ, Burham RD and Scifres DR
    1982 Appl. Phys. lett. 41 317
Capasso F, Tsang WT, Hutchinson AL and Williams GF 1982 Appl. Phys.
    lett 40 38
Cappy C, Vernaeyen C, Vanoverschelde A, Salmer G, Delagebeaudeuf D,
    Linh NT and Laviron M 1982 GaAs IC Symp. New Orleans, November
Chen CY, Cho AY, Garbinski PA and Bethea CG 1981 IEEE Electron Dev.
    Lett. EDL2 290
Delagebeaudeuf D, Delescluse P, Etienne P, Laviron M, Chaplart J and
    Linh NT 1980 Electron. Lett. 16 667
Delagebeaudeuf D, and Linh NT 1982a IEEE Trans. Electron Dev. ED29 955
Delagebeaudeuf D, Delescluse P, Laviron M, Tung PN, Chaplart J,
    Chevrier J and Linh NT 1982 b Inst. Phys. Conf. Ser. N° 65 Chap. 5
    pp 393-398
Döhler GH 1972 Phys. Stat. Solid (b) 52 533
Döhler GH 1979 J. Vac. Sci. Technol. 16 851
Dingle R, Wiegmann Wand Henry C 1974 Phys. Rev. Lett. 33 827
Dingle R 1975 Festkörperprobleme XV Advances in Solid State Phys.
    Pergamon pp 21-48
Dingle R, Störmer HL, Gossard AC and Wiegmann W 1978 Appl. Phys. Lett.,
    33 665
Drummond TJ, Kopp W, Morkoç H and Keever M 1982 Appl. Phys. Lett. 41 277
Esaki L and Tsu R 1969 IBM Res. Int. Report RC 2418 March 26
Fishmann G and Calecki D 1983 Physica 117B & 118B North Holland p. 744
Hersee SD, Baldy M, Assenat P, de Cremoux B and Duchemin 1982a Electron.
    Lett. 18 870
Hersee SD, Hirtz JP, Baldy M and Duchemin JP 1982b Electron. Lett. 18
    1077

Hess K, Morkoç H, Schichijo H and Streetman BG 1979 Appl. Phys. Lett.
35 469
Hess K, Vojak BA, Holonyat N, Chin R and Dapkus PD 1980 Solid State
Electron. 23 585
Hiyamizu S 1982 2nd Internat. Symp. MBE and CSI Tokyo, August
Hwang JCM, Temkin H, Kastalsky A, Störmer HL and Keramides VG 1982 MBE
Workshop
Joshin K, Mimura T, Niori N, Yamashita Y, Kosemura K and Saito J 1983
MTT Boston, June
Keever M, Schichijo H, Hess K, Banerjee S, Witkowski L, Morkoç H and
Streetman BG 1981 Appl. Phys. Lett. 38 36
Kiehl Ra, Feuer MD, Hendel RH, Hwang JCM, Keramidas VG, Allyn CL and
Dingle R 1983 41st Annual Device Research Conf. Burlington, June
Laviron M, Delagebeaudeuf D, Delescluse P, Chaplart J and Linh NT 1981
Electron. Lett. 17 536
Linh NT 1983a Spring Meeting of the German Physical Soc. Freudenstadt, to
be published in Festkörperprobleme XXIII
Linh NT 1983b 8th European Specialist Workshop on Active Microwave
Semiconductor Devices, Maidenhead, May
Linh NT, Laviron M, Delescluse P, Tung PN, Delagebeaudeuf D, Diamand F.
and Chevrier J. 1983 Cornell Conference, August
Lee CP, Miller DL, Hou D and Anderson RJ 1983 41st Annual Device Research
Conf. Burlington, June
Malik RJ, Board K, Eastman LF, Wood CEC, Aucoin R and Ross RL 1980 Inst.
Phys. Conf. Ser. N° 56 Chap. 9 pp 697-710
Mimura T, Hiyamizu S, Fuji T and Nanbu K 1980 Jpn J. Appl. Phys. 19 L225
Mimura T, Joshin K, Hiyamizu S, Hirosaka K and Abe M 1981 Jpn J. Appl.
Phys. 20 L598
Mudares Al and Foulds KWH 1983 8th European Specialist Workshop on Active
Microwave Semiconductor Devices, Maidenhead, May
Mori S and Ando T 1980 Surface Sci. 98 101
Niori M, Saito T, Joshin K, Hiyamizu H, Hikosaka K and Abe M 1981 Jpn
J. Appl. Phys. 20 L598
Nishiuchi K, Mimura T, Kuroda S, Hiyamizu S, Nishi H and Abe M 1983
41st Annual Device Research Conf. Burlington, June
Osbourn GC 1982 J; Appl. Phys. 53 1586
Tsang WT 1981 Appl. Phys. Lett. 39 786
Tsang WT, Logan Ra, Ditzenberger JA 1982 Electron. Lett. 18 845
Tsui DC, Gossand AC, Field BF, Cage ME and Dzuiba RF 1982a Phys. Rev.
Lett. 48 3
Tsui DC, Störmer HL, Gossard AC 1982b Phys. Rev. Lett. 48 1559
Tung PN, Delescluse P. , Delagebeaudeuf D, Laviron M, Chaplart J and
Linh NT 1982 Electron. lett. 18 517
Tung PN, Delescluse P and Linh NT 1983 unpublished.

*Inst. Phys. Conf. Ser. No. 69*
*Paper presented at ESSDERC/SSSDT 1983, Canterbury 13–16 Sept. 1983*

29

# Diffraction and interference optics for monitoring fine dimensions in device manufacture

H.P. Kleinknecht

Laboratories RCA Ltd., Zurich, Switzerland

Abstract. A review is given of the use of interference for layer thickness monitoring, in particular during epitaxial growth, sputter deposition, plasma etching and wet etching using grating test patterns. We also describe a thin film tester using white-light interference and an optical profilometer for power device wafers. For measurement of lateral dimensions, i.e. linewidths in LSI, we have developed a diffraction technique using gratings on masks and wafers.

## 1. Introduction

The application of interference and diffraction optics in semiconductor device fabrication is a very large field, which can not be covered exhaustively in a one-hour talk. For this reason, I must be very selective, and I have decided to be selective in a very personal way: I will try to concentrate on pertinent work which has been done in our laboratory of RCA in Zürich, Switzerland. I think that the number of projects, I will be able to talk about, is nevertheless large enough to give a representative view of the possibilities of optics in device manufacture.

At first I will talk about thickness monitoring by interference in general, and I will then describe as examples an epitaxial-growth monitor, a film deposition monitor, a double-beam plasma etch monitor and a grating etch monitor. Next I will describe a thin film thickness tester and an optical profilometer. Finally, we will deal with lateral dimensional control in the form of linewidth measurement using diffraction grating test patterns.

## 2. Thickness Monitoring During Deposition or Etching of Layers

Before describing the various set-ups for thickness monitoring, it is of advantage to look at Fig. 1 as a reminder of the general principle of interference. Figure 1a is a sketch of a light beam considered at near normal incidence as it is reflected and transmitted many times at the thin, transparent layer of refractive index, $n$, and thickness, $D$, which is deposited on a substrate with a different refractive index, $n_S$. We also give the formula for the reflected intensity, $I_R$; $r_1$ and $r_2$ are the amplitude reflectivities at the surface and at the interface, respectively (Born and Wolf (1965)). $\emptyset$ is the phase difference between consecutively reflected beams, which is proportional to the thickness, $D$, and the refractive index, $n$, and inversely proportional to the wavelength, $\lambda$. Due to the superposition of the amplitudes of all the reflected partial beams, one gets the total reflected intensity as a periodic function of $\emptyset$. In Fig. 1b the reflected intensity is plotted as a function of $D$ for $SiO_2$

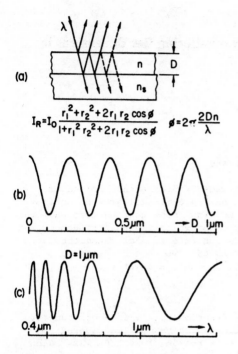

$$I_R = I_0 \frac{r_1^2 + r_2^2 + 2r_1 r_2 \cos\phi}{1 + r_1^2 r_2^2 + 2r_1 r_2 \cos\phi} \qquad \phi = 2\pi \frac{2Dn}{\lambda}$$

Fig. 1. The principle of interference.   Curves (b) and (c) are
for $\lambda$ = 0.6328 $\mu$m, n = 1.46 (SiO$_2$)
and n$_s$ = 3.5 (Si) :

on Si, and in Fig. 1c we have
plotted the intensity as a function
of the wavelength, which goes inversely with $\phi$.  Oblique incidence
does not change this situation much.
It adds only a cosine of the angle
of incidence as a factor to the
denominator of $\phi$.

Figure 1b teaches us how to measure
and monitor the change of the layer
thickness, i.e. the etch rate or
growth rate, by following the
reflection of a monochromatic light
beam, such as a laser beam, and
Fig. 1c teaches how a fixed layer
thickness can be measured by using
light of changing wavelength.  We
will use both possibilities in the
following applications.  The pre-
requisites for this to work is that
the layer to be measured is sufficiently transparent to the probing
beam and that the substrate has a
refractive index different from that
of the layer to give sufficient
interface reflectance.  Multiple
layers and specular reflection from
the bottom surface of the substrate
require a more elaborate analysis.

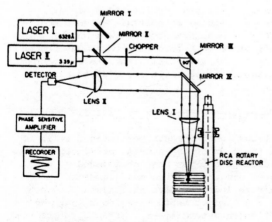

Fig. 2. Growth rate monitor for epitaxial growth of Si.

Figure 2 shows a first application of this principle in
the form of a growth rate
monitor developed by
G. Harbeke (see also Dumin
(1967)) for use in conjunction with a rotational disk
epitaxial reactor.  Here a
stack of silicon wafers is
rotated around a vertical axis
while the gas for the gas-
phase deposition is supplied
by nozles at the circumfer-
ence of the wafers (not
shown in the figure).  The
beam of the He-Ne laser II,
operating at 3.39 $\mu$m, is
directed through quartz optics
and through a planar quartz
window in the reactor envelope

at the growing layer.  This particular wavelength is close to the minimum
of absorption (between lattice and free-carrier absorption) at the growth
temperature.  The return beam is deflected by a beam splitting mirror and
focussed on the photo-detector.  The use of a chopper, a phase-sensitive

amplifier and a recorder is straightforward. Laser I is a red laser used for alignment of the infrared beam. The top two wafers are dummies with holes for passing the beam. They are used to control the temperature end losses. The third wafer, on which one does the measurement, is also not necessarily a product wafer of Si but has a sapphire substrate in order to get the necessary interface reflection. Figure 3 gives a recording of the signal during an epitaxial deposition run. This recording corresponds to Fig. 1b, if the time is correlated with the incrreasing layer thickness. Counting the oscillations gives thickness and growth rate at any time during the process. The amplitude decrease with growing layer thickness is due to the non-zero absorption and limits the technique in this case to 16 μm, which depending on the test wafer position in the reactor can correspond to much thicker layers on the Si process wafers.

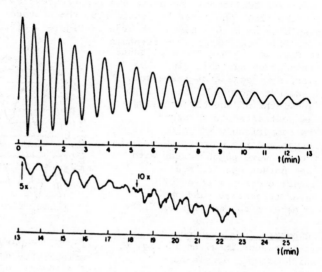

Fig. 3. Recording of the epitaxial growth rate monitor. Separation between adjacent maxima corresponds to 0.498 μm.

This epitaxial growth monitor can use regular lens and mirror optics because of the special geometry of the rotary disk reactor. The situation is more complicated in Fig. 4., which shows a film deposition monitor for multiple-layer sputter deposition by M.T. Gale, H.W. Lehmann, E. Heeb and K. Frick (1982). Here the substrate has to be moved on a table from one source position to the other, and the measurement is facilitated by flexible fiber optics. The light sources are two light-emitting diodes (LED) of two colors, red and green, which

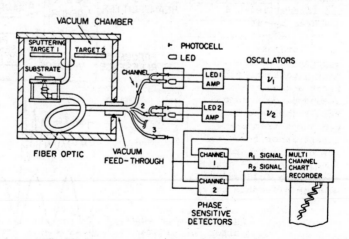

Fig. 4. Block diagram of the deposition monitor.

couple into two partial bundles of the fiber optics and illuminate the growing layers through the transparent substrate. A small percentage of the primary light is fed back to two reference photo-diodes which are used to stabilize the LED outputs. The light of both colors reflected from the growing layers is collected by a third bundle of the fiber optics and guided back out to a Si photocell. Since the red and green LED's are pulsed with different frequencies, $\nu_1$ and $\nu_2$, the phase-sensitive amplifiers locked to the two frequencies can separately filter out the "red" and "green" signals from the photocell, which then can be displayed on a two-channel recorder. Figure 5 gives the recording for a

TiO$_2$ film, which again corresponds to Fig. 1b. Again the amplitudes are decreasing, this time, however, not because of the absorption, but because of the finite bandwidths of the LED emission (as contrasted to a laser). The comparison with calculated curves and the comparison of the two wavelength curves with one another enables the accurate determination of layer thicknesses and growth rates also for multilayer films.

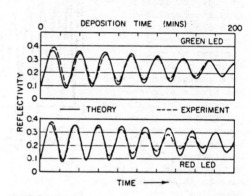

Fig. 5. Experimental and theoretical reflectivity curves for the deposition of a 600 nm thick TiO$_2$ film.

An example for etch control by interference is given in Fig. 6, which shows a double-beam monitor in a plasma etcher built by H.W. Lehmann. The beam of a red He-Ne laser is split into two partial beams which are reflected from two different areas, A and B, of the sample, and the reflected intensities are measured by two photodiodes. The multivibrator alternates between making contact to the first and to the second diode. As a result the recorder pen jumps from one interference trace to the other. The top envelope of this composite trace follows the etch interference curve

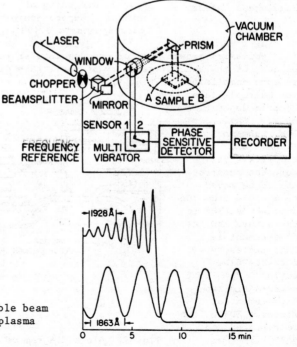

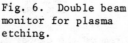

Fig. 6. Double beam monitor for plasma etching.

of the layer A, which is a polysilicon layer on sapphire in this case. Here the amplitudes of the oscillations increase with time, because the layer gets thinner, and finally at zero thickness the oscillations stop. The other envelope shows the etching of the area B, which is covered with photoresist giving slower and more uniform oscillations. Hence, this set-up gives an end point detection, and it gives an accurate comparison of the etch rates of polysilicon and photoresist.

## 3. Determination of Etch Rate and Undercutting Rate by the Use of Grating Test Patterns

The above described three examples of layer thickness monitoring by interference can be used only if the area illuminated by the probing light beam is uniform over the beam cross-section.
If the beam simultaneously strikes two or more areas of different thickness or different etch rates, the reflected intensity can no longer be interpreted in the way described above. This can be a problem when patterned integrated circuit wafers have to be monitored during etching. On these wafers large uniform areas can not be wasted and it may be difficult to focus the test beam on the small available uniform areas during the etching process.

One way out is to provide a small test pattern on the wafer in the form of a diffraction grating (Kleinknecht and Meier (1978)). This is sketched in Fig. 7. The incident laser beam is reflected in the form of a diffraction pattern consisting of a zero-order and first and higher orders with fixed diffraction angles, $\varphi_1$, $\varphi_2$, etc., given by the grating constant, d. As the area in the spaces between the photoresist lines is etched, the intensities of these orders will go through oscillations similar to those for planar interference. But one can now place the photosensor at the angular position of the first order (not the zero-order or the specular reflection), and the laser beam can now be larger than the grating area and can also be allowed to strike non-uniform circuit parts on the wafer,

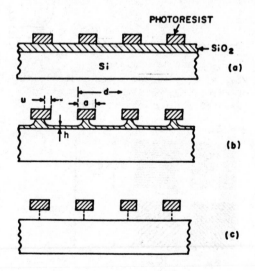

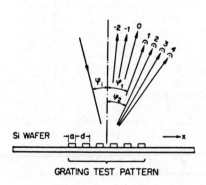

Fig. 7. Diffraction by the grating test pattern on a wafer

Fig. 8. Etching of a grating test pattern

without disturbance of the first-order beam, which comes exclusively from the grating area. Figure 8 shows cross-sections of such a grating test pattern, unetched (a) and partially etched (b). The grating test pattern consists of photoresist lines, which can be produced in the same step used to delineate the etching process on the circuit part of the wafer. One merely has to include the grating pattern on a test chip on the particular mask level, and the grating can be as small as 0.5 mm in diameter. Figure 8b does not only indicate the normal etching, which thins the $SiO_2$ and produces the oscillations in the reflected first-order intensity. It also shows the lateral underetching, and in Fig. 8c the underetching has reached completion, i.e. it has progressed to half the width of the grating lines. This, in a liquid etch, will allow the grating bars to shift and to fall off, which destroys the grating periodicity and causes the measured first order intensity to drop. Figure 9 shows recorder traces of three such etching experiments, again $SiO_2$ on Si with a photoresist grating on top etched in hydrofluoric acid. At first we have the interference oscillations from which the vertical etch rate can be calculated, and then, after about 6.5 minutes, the sharp drop indicates that the lateral etching has gone as far as half the width of the grating lines.

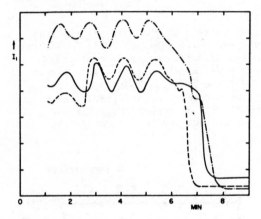

Fig. 9. First-order intensities vs. time of three samples with 10 um gratings measured during wet etching.

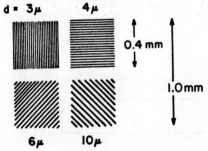

Fig. 10. Schematic drawing of the four-grating test patterns

Figure 10 shows a more complex test pattern consisting of four gratings. The grating periods and the orientation of the grating lines are different on all four gratings. Therefore, if the whole pattern is illuminated by a laser beam, the first orders of all four gratings are reflected in different directions and can be measured simultaneously and separately by four properly placed photocells. The linewidth of each grating is half the period, i.e. 1.5, 2, 3 and 5 $\mu$m, respectively. Consequently, the four signals of the four photocells will during the etching process first oscillate simultaneously up and down and will then drop one by one at 0.75, 1, 1.5 and 2.5 $\mu$m of undercutting, respectively. This is exactly what can be seen on the 4-channel trace of Fig. 11. The recording simultaneously allows the monitoring of normal as well as lateral etch rate.

This type of monitoring was done during wet etching. We also have applied the grating technique to the monitoring of plasma etching, in which case, however, the undercutting, i.e.

the falling off of the lines, can not be observed.

## 4. A Thin-Film Thickness Tester based on White-Light Interference

In contrast to all of the techniques described so far, which operate with light of one or two wavelengths, white light is used in a thickness tester invented and designed by J.R. Sandercock (1983), which is accurate, simple, compact and relatively cheap. It has widespread use within the semiconductor industry for quick measurement of layers of $SiO_2$ on Si, $Si_3N_4$ on Si, Si on sapphire and of photoresist on chrome and on Si. The thicknesses, which can be measured, range from 0.5 to 10 $\mu$m, and the accuracy is about 1%.

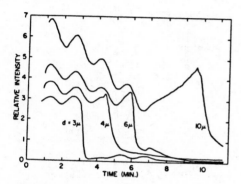

Fig. 11. Recorder trace of a four-grating etch run.

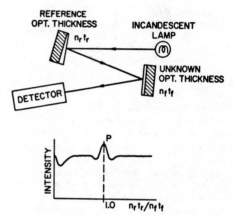

Fig. 12. Schematic of the thin film tester and a plot of intensity at the detector vs. optical thickness ratio of reference and unknown sample.

Figure 12 shows its principle of operation. The light from an incandescent lamp is first reflected from a reference sample of known film thickness and then from the unknown film. The intensity of the twice reflected beam, as measured with a broad band pyroelectric detector, is a maximum, if the optical thicknesses of both layers, i.e. the products of refractive index and thickness, are equal. This intensity vs. the optical thickness ratio is plotted in the lower part of the figure. The explanation for the maximum can be obtained by looking at Fig. 1c, which shows that at the first sample only certain wavelengths are strongly reflected while others are suppressed. And only if the optical layer thickness of the second sample is the same, the same wavelengths will predominantly be reflected at the second sample resulting in high intensity at the detector. If the two optical thicknesses are not equal, the two peak structures do not coincide, and the intensity at the detector will be lower. There is only one unique maximum in Fig. 12, because the peaks along the $\lambda$-axis in Fig. 1c are not uniformly spaced. As indicated in Fig. 12, the known reference sample has a wedge shaped layer. Hence, by laterally shifting this wedge in the beam, its effective thickness can be changed, while the intensity is monitored. Then the position, where the maximum occurs, corresponds to the thickness of the unknown layer.

In practice not a simple rectangular wedge is used but a semicircular wedge, i.e. one half of an oxidized Si wafer, where the $SiO_2$ thickness increases linearly with rotation angle from zero to 10 $\mu$m as indicated in Fig. 13.   The semicircular $SiO_2$ wedge is produced by etching in buffered HF by immersing the disc slowly into the etch liquid such that the various sectors of the wafer are exposed to the etching process for times varying lienarly with the angular position of the wafer.   After etching the wafer is attached to a circular disc (Fig. 13) and mounted on a motor shaft. Figure 14 gives more details of the instrument such as lenses and mirrors

Fig. 13. The reference wedge.

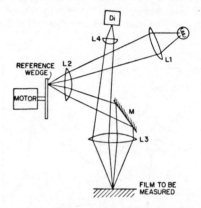

for focussing the light in succession on the rotating wedge, the unknown sample and the broad-band detector.   During the measure-ment the disc is rotated with 1 revolution per second, which is the repetition frequency of the

Fig. 14. Detailed design of the optics of the thin film tester

measurement.   Figure 15 gives the waveform of the intensity at the detec-tor, which corresponds approximately to the curve of Fig. 12.

Suitable circuitry is used to sense the maximum of the intensity waveform as a function of time.   The position of the maximum in the time regime can be cali-brated in terms of the optical thick-ness and, with a known refractive index, in terms of the geometrical layer thickness of the unknown sample.

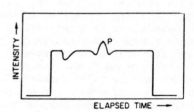

Fig. 15. Intensity at the de-tector vs. time.

## 5. An Optical Profilometer for Surface Contours of Si Power Device Wafers

The fabrication of power transistors and thyristors involves the etching of deep grooves (30-100 $\mu$m) into the Si wafers.   These grooves are the site where the high collector field meets the surface and where the passivating glass and oxide layers have to be applied (Fig. 16).   There-fore, the groove depths have to be measured in manufacturing to +5%, and this has to be done without touching the wafer surface.   Since these dimensions are large compared with the wavelength of light and since the etched surfaces can be rough, the above described interference techniques can not be used.   We have for this application designed and built an optical profilometer (Kleinknecht and Meier (1983)).   It essentially con-sists of a high-power microscope equipped with a laser attachment which automatically steers and holds the microscope focussed to the sample surface . An electronic stylus type depth gauge, a "linearly variable differential transducer" (LVDT) resting on the shoulder of the microscope measures the vertical excursions of the microscope, which follow the sample contour. This instrument is fast and accurate to +1 $\mu$m with a range of well over 100 $\mu$m. It is in use at present in RCA's power device plant in Pennsylvania.

Figure 17 shows the essential elements of the profilometer: the microscope comprising ocular and objective lens and two beam splitters for the coupling in and out of the laser beam. The laser beam is brought to a first focus, "1", which is imaged by the objective lens to the sample sur-

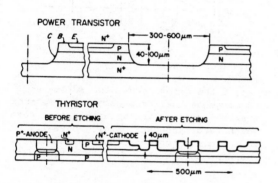

Fig. 16. Cross-sections of Si power device wafers.

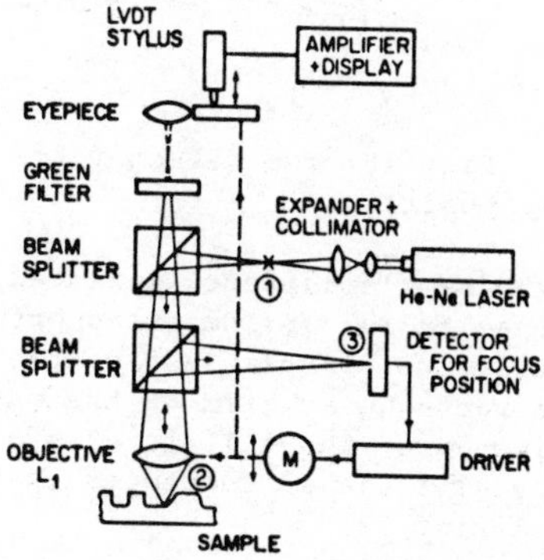

Fig. 17. Principle of the optical profilometer

face at "2", reflected back again through the objective and re-focussed at "3". The longitudinal position of focus "3" (left-right direction on Fig. 17) is strongly dependent on the sample surface height. The box called "Detector for Focus Position" in Fig. 17 is described below. It produces a positive or negative d.c. voltage depending on whether the focus "3" is too far to the right (sample surface too close to the objective) or too far to the left (sample surface too far), respectively. This voltage drives a servo motor which raises or lowers the microscope such that the laser focus "2" (and the visual image as seen through the eye piece) is kept automatically at the surface. The sample is in addition illuminated by white light so that the sample surface can be observed visually through the ocular lens. The green-filter reduces the laser radiation entering the observer's eye to a safe level. For a LVDT we use a commercial instrument (SILVAC) which solves the calibration, accuracy and display problem for us.

The contents of the box "Detector for Focus Position" are shown in Fig. 18, essentially a third beam splitter, which deflects one half of the beam to an aperture with a photocell (sensor 1) before it comes to a focus. The other half goes straight to the focus and behind that reaches a second aperture and sensor 2. The unfolded beams are shown at the bottom of Fig. 18 indicating that sensor 2 gets more light than sensor 1, if the focus goes to the right and vice versa. The difference signal, $P_2-P_1$, is suitable for driving the servo motor. The top of Fig. 19 is an unfolded view of the total optics in a simplified way: the incoming beam, the objective lens, the sample, the reflected beam and the apertures (assumed to be circular holes). The figure demonstrates a serious problem: If

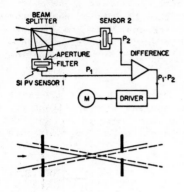

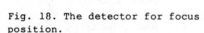

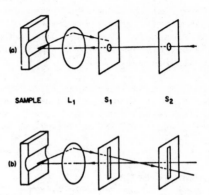

Fig. 18. The detector for focus position.

Fig. 19. The tilt problem: slit apertures.

the sample surface is tilted, such as at the groove edge, the reflected beam falls outside the apertures and disables the focus detector. Fortunately, we can make use of the fact that the grooves in our application form a rectangular grid on the wafers, so that for the horizontal grooves for instance, the tilt is always vertical. This situation can be accommodated by slit shaped apertures in the focus detector as shown in the bottom part of Fig. 19.

Fig. 20 is a block diagram of the whole instrument, which is mostly self-explanatory. The "sum" signal is used with a "comparator" to switch off the feed-back loop, if the signal is low (at the groove edges). The loop is also switched off automatically, if the microscope is too far out of focus. The display can be zeroed to facilitate the direct reading of height differences. Figure 21 gives some measurement traces across a Si surface into which rather rough steps have been cut with a diamond saw. Using the lower magnification objectives, 16x and 25x, one gets traces

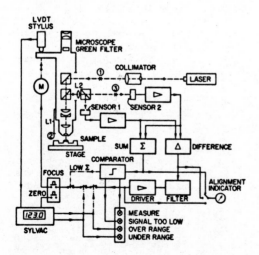

Fig. 20. Block diagram of the optical profilometer.

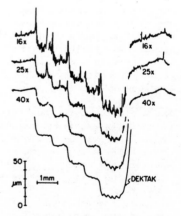

Fig. 21. Scan across steps on a Si sample.

which are somewhat noisy. But the trace with 40x corresponds very well with the trace from a mechanical stylus of a "Dektak", which scraped directly along the Si surface. Needless to say, in the factory the Dektak would be much too slow, too delicate and would damage and contaminate the wafer surface. This is, of course, in contrast to the optical profilometer.

## 6. Linewidth Monitoring using Diffraction Grating Test Patterns

Up to now we have mainly been concerned with measuring and monitoring of the dimensions normal to the major plane of the wafers and the devices. We will now turn to the measurement of the lateral dimensions, in particular of the very fine features on VLSI circuits, such as source-drain spacings and the width of conducting stripes, which are the result of the control of linewidth in mask making, photoresist processing and etching. For wide lines the preferred tool for this is the optical microscope. However, as the linewidths are approaching dimensions of microns and below, the microscope images are distorted by diffraction, the evaluation of which requires knowledge of the line shape and elaborate analysis of the image and/or calibration, which is what is done in a number of commercial linewidth testers.

We have approached the problem from the other direction by using the diffraction effect directly for linewidth determination instead of trying to eliminate it or correct for it (Kleinknecht and Meier (1980)). This means that we do not make the measurement on individual lines but on periodic arrays of lines, i.e. diffraction gratings. In other words we are not measuring the linewidths directly on the circuit area of the mask or wafer, but we indlude on the mask at the edge or in test chip areas small grating test patterns (0.5 mm x 0.5 mm) with grating lines of the same widths as the critical dimensions on the circuit area. These same test patterns are in the normal photoresist process transferred from the mask to the wafer along with the circuit patterns and can then also be used for testing of linewidths on the wafers. By measuring the linewidths on the grating and noting their deviations from the design value, we also know the deviation in linewidth on the product areas of the same mask or wafer.

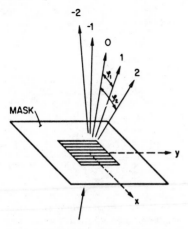

Fig. 22. Diffraction by a grating test pattern on a mask

Figure 22 shows such a grating test pattern on a mask, as it is probed in transmission by a laser beam coming from below. The transmitted light is split up into the well-known diffraction orders. The same thing happens when a grating is probed in reflection, such as a test pattern on a wafer (see Fig. 7). The diffraction angles, $\varphi_1$, $\varphi_2$, etc., depend only on the laser wavelength and the grating period, which is fixed by the test pattern design. The intensities, however, of the various orders strongly depend on the details of the structure and the width of the lines. Hence, the measurement of the intensities of the various orders can be used for linewidth determination.

For linewidths larger than the wavelength (in our case 0.633 μm of the red He-Ne laser) one can with good approximation use the scalar theory of Kirchhoff and Fraunhofer (see Born and Wolf (1965)) and for the testing of masks one can concentrate on the simple case of a rectangular grating profile, such as shown in Fig. 23, where both the modulus and the phase of

the transmittance or reflectance are rectangular step functions of the coordinate along the main plane of the grating. In this case the ratio of the second to the first order intensity is simply $I_2/I_1 = \cos^2(\pi a/d)$, where a is the linewidth and d is the grating period. Note that for this simple case, and only for this case, the thickness and the refractive index or reflectivity of the lines do not enter at all. Hence, knowing d, the linewidth a can be determined by simply measuring the intensities of the second and first orders in relative terms. The cosine-square formula also teaches that this measurement procedure will be most accurate around a = d/4, which gives us the design recipe for the test pattern.

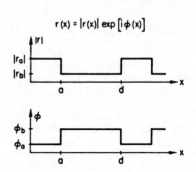

$$r(x) = |r(x)| \exp\left[i\,\phi(x)\right]$$

Fig. 23. Rectangular transmittance or reflectance function.

We have made extensive tests on the accuracy of the grating technique for masks by comparing results of the laser diffraction with linewidth measurements taken from scanning electron microscope pictures (SEM) of the same gratings. Because the grating period is known, the SEM measurement involves only the determination of a/d of the gratings, which does not require any knowledge of the SEM magnification. A typical set of measurements is given for about 40 different test gratings on an electron beam fabricated mask in Fig. 24. The top of the figure gives the direct relation between laser and SEM measurements

for 10 gratings. The bottom plots the relative deviations between laser and SEM data for all 40 gratings as a function of linewidth. The error is typically 5%. In Fig. 25 we compare these relative deviations from the SEM data (for another chrome mask) with two optical microscope measurements, one with an image splitting eye piece, the other with a commercial TV-electronic system based on an optical microscope. The comparison shows that for linewidths above 5 μm the errors are about equal for all three systems (if the SEM is considered to be the standard). But below 5 μm the laser diffraction technique is clearly more accurate. Sometimes the interference effect tue to the multiple reflections of the laser beam at the glass-air surface of the mask substrate can cause an error. This problem can be solved by using Brewster's Angle

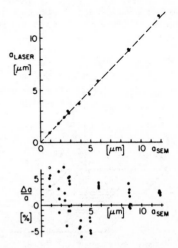

Fig. 24. Comparison of laser diffraction and SEM measurements for 40 gratings on a chrome mask.

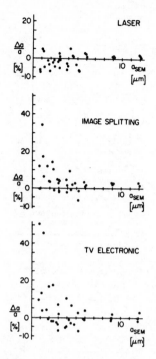

conditions for the laser beam, which eliminates the glass-air reflections. Tests with very fine gratings showed that for lines below 0.6 μm larger errors appear, as expected due to the breakdown of the scalar diffraction theory.

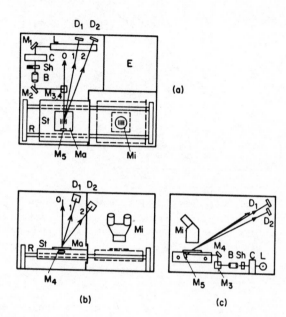

Fig. 25. Comparison of diffraction, microscopic and SEM measurements for a chrome mask.

Fig. 26. Schematic view of a mask diffracto-meter: (a) top, (b) front, (c) side.

Figure 26 is a simplified drawing of an optical diffractometer based on this principle, which was built for testing masks. The beam of the He-Ne laser, L, goes through a chopper, C, a shutter, Sh, and a beam collimator, B. The mirror, $M_5$, directs the beam at Brewster's Angle incidence through the grating test pattern on the photomask, Ma. The first and second orders are intersected by the Si photodiodes, $D_1$ and $D_2$, which are posi-tioned by a microprocessor controlled mechanism according to the grating period, which is dialed in. The microprocessor converts the two inten-sities (from two phase-sensitive amplifiers) according to the cosine-square formula into the linewidth. For positioning the test pattern properly in the beam, the mask holder with the x-y stage, St, is prior to the measurement adjusted under a microscope. Then it is automatically transferred into the optical compartment to the correct position in the laser beam, and the testing process is initiated. For the testing of wafers a similar diffractometer was constructed (Bösenberg and Kleinknecht (1982/3)), which had to work in reflection and had to have provisions for rotational adjustment of the wafers in addition to x-y. Also two choices of incidence angles, near normal and Brewster's, were provided. The wafer measurement is, however, not as straightforward as the one on masks, be-cause the lines on wafers are photoresist, etched oxide, poly or metal with line shapes, which may not be as rectangular as the chrome lines on masks (Fig. 23). As a result one must calibrate the diffractometer line-width measurements on wafers for any processing type with the SEM. Fig. 27 and 28 give data for many kinds of grating lines, which show that the

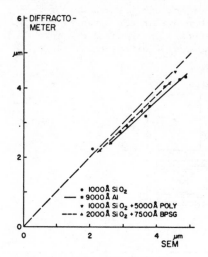

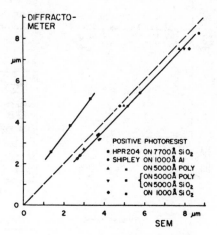

Fig. 27. Diffractometer vs SEM data for gratings etched into various types and combinations of layers on Si wafers.

Fig. 28. Diffractometer vs SEM data for gratings developed in photoresist on various layers on Si wafers (no etching)

precision and the correlation with SEM data is sufficiently good for such a calibration procedure to be possible. Once this calibration is established, the diffractometer measurements can also be extended to linewidths below 0.6 μm, since, with SEM calibration, we do not rely on the cosine-square formula any more. This is expected to become valuable for sub-micron structures.

In the above described linewidth monitoring the grating test patterns were very small and could fit on product wafers or on masks for making product. We have also done experiments with test masks and wafers, which contained nothing but one large grating over the whole surface with 8 μm period and nominally 2 μm linewidth. When these masks or wafers were scanned through a laser beam, the continuous measurement and plotting of second- to first-order intensity ratio gave an image of linewidth non-uniformity across the wafer area due to processing. (Kleinknecht, Meier and Ham (1980) and Ham, Kleinknecht and Meier (1980)). The experimental apparatus is shown in Fig. 29. The horizontal scan motor pushes the wafer through the beam and at the same time causes the recorder pen to draw a horizontal line. After each horizontal scan the y-position of the wafer and of the pen are stepped up. Simultaneously, the signal $I_2/I_1$ from the sensors $S_1$ and $S_2$ is added to the y-input of the recorder. Hence, the vertical deviations of the recorder from the horizontal lines are a measure for the local linewidth deviations on the wafer. The result is a map of linewidth of the whole wafer area, which can serve as a tool to pinpoint deficiencies in the mask or wafer processing. Figure 30 is such a map of a mask. The right hand side scale gives the coordinate on the mask. At the far left is a micron scale for the linewidth deviations. As can be seen, the non-uniformity, as indicated by the nearly perfectly horizontal traces, is only about 0.1 μm. Figure 31 shows the map of a wafer with photoresist lines. The linewidth non-uniformity here is larger and seems to follow a concentric circular pattern. Figure 32 shows

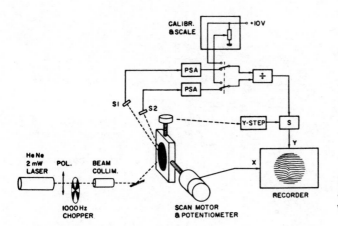

Fig. 29. Experimental apparatus for line-width scanning.

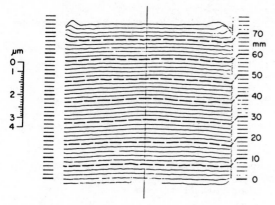

Fig. 30. Linewidth plot of a mask.

a well-defined semicircular band of linewidth deviation on another wafer. The radius of this band coincides with that of the slit aperture of the Perkin-Elmer aligner, which was used for this process. The band is caused by a non-uniform, jerky movement of the aperture across the wafer during exposure. This is a well-known failure possibility of this type of aligner.

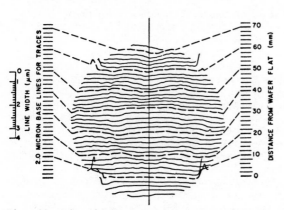

Fig. 31. Linewidth plot of a wafer with photoresist lines.

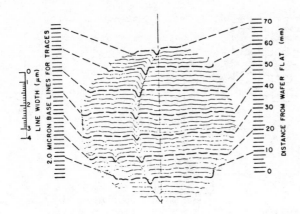

Fig. 32. Linewidth plot of a wafer showing "banding" due to jerky motion of the P.E. aligner aperture.

## 7. Conclusion

As said in the introduction, this review was not intended to be comprehensive in any way. It is hoped, however, that the examples given here help to stimulate the application of optics for process control in semiconductor device manufacture. We have seen that in most examples here the simplicity and monochromaticity of lasers was utilized. As a consequence, there is in all factory applications a problem of laser safety, which has to be taken care of by interlocks, shutters, shields and filters - and a lot of red tape. On the other hand, in many cases, even in our examples, one could have used incoherent light sources such as LED's

The author would like to acknowledge the support of many of the members of this laboratory, mostly the authors of the cited references, who have supplied material and suggestions for this review.

## References

Born M and Wolf E  1965  Principles of Optics, (London: Pergamon)
Bösenberg W A and Kleinknecht H P 1982 Solid State Technol.  Oct. pp 110
    (Part I)
Bösenberg W A and Kleinknecht H P 1983, dito August (PartII)
Dumin D J 1967 Rev. Scient. Instr. 38 1107
Gale M T, Lehmann H W, Heeb E and Frick K 1982 J. Vac. Sci. Technol.
    20, 16
Ham W E, Kleinknecht H P and Meier H 1980 Europhysics Conference Abstracts
    4H 205, ESSDERC (York)
Kleinknecht H P and Meier H 1978 J. Electrochem. Soc. 125 798
Kleinknecht H P and Meier H 1980 Applied Optics 19 525
Kleinknecht H P, Meier H and Ham W E 1980 Europhysics Conference Abstracts
    4H 203  ESSDERC (York)
Kleinknecht H P and Meier H 1983, SPIE Proceedings 398 38 (international
    Technical Conference/Europe - Geneva)
Sandercock J R 1983 to be published

*Inst. Phys. Conf. Ser. No. 69*
*Paper presented at ESSDERC/SSSDT 1983, Canterbury 13–16 Sept. 1983*

# Hot electron diodes and transistors

J.M. Shannon

Philips Research Laboratories, Redhill, Surrey, England.

Abstract   Hot electrons play a major role in the behaviour of many
semiconducting devices and a variety of fascinating devices have been
proposed which use them.  This paper concentrates on the properties and
performance of hot electron diodes made in the bulk of a semiconductor
both as unipolar structures and internal hot electron thermionic
emitters.  The nanometre technology required to make some of the most
interesting devices is outlined and it is shown that hot electron
transistor structures can be realised with both voltage and current
control to give transistors with performances which are potentially
superior to existing devices.

## 1. Introduction

Hot electrons are used in a number of devices, particularly two terminal
microwave devices (Carroll 1970) where high internal fields are used to
promote avalanche breakdown or intervalley scattering both of which can
lead to negative resistance and current oscillations (Ridley 1977).  In
this paper we shall be concerned with electric fields inside a
semiconductor which are considerably in excess of the avalanche breakdown
field in bulk material.  Large electric fields occur naturally between
highly doped n and p regions in a semiconductor and provided the kinetic
energy gained by a carrier as it passes between these regions is not large
compared with the energy band gap there is insufficient energy to generate
a hole-electron pair.  Structures can therefore be designed in which the
internal electric fields are an order of magnitude greater than the bulk
breakdown field, and band bending occurs over very short distances
comparable with the electron mean free path in the semiconductor.

Hot electron diodes require the formation of high electric fields in
structures capable of withstanding an applied voltage.  The first of these
was the camel diode (Shannon 1979a) which is one of a category of devices
that we shall call bulk unipolar diodes (Board 1982).  Included in this
category are planar doped barriers (Malik et al 1980, Malik et al 1981)
and bulk barrier diodes (Mader 1982).  In all cases potential barriers are
formed in a semiconductor in thermal equilibrium due to the presence of a
narrow plane of ionised charge.  Some of these structures are more
suitable as hot electron emitters than others and in the first section of
this paper criteria which need to be satisfied are outlined together with
some measurements on camel diode structures made using ion implantation or
molecular beam epitaxy.

The formation of nanometre structures required for hot electron devices is described in section 3 and their incorporation into gate controlled and monolithic hot electron transistor structures forms the basis of sections 4 and 5.

## 2. Hot Electron Diodes

The metal-semiconductor Schottky diode is frequently referred to as a hot electron or less specifically a hot carrier diode (Carroll 1974), because carriers which pass over the barrier from the semiconductor enter the metal at energies very much greater than the mean energy of the free electrons in the metal (Fig. 1). Carriers that pass from the metal into

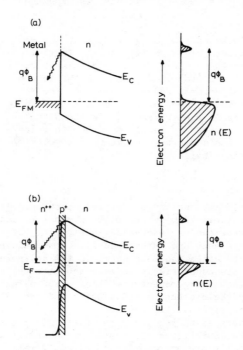

Fig. 1    Band shapes for a Schottky diode and a bulk unipolar camel diode. Also shown schematically is the distribution of electrons in the neutral metal (a) or semiconductor (b) at the edge of the barrier.

the semiconductor also have a high energy in relation to the carriers in the neutral semiconductor but the comparatively small electric field in the semiconductor and the extent of the space charge region enables the carriers to thermalise as they move down the potential gradient of the barrier. The ability of Schottky barriers to inject hot electrons into a metal, however, has not been exploited although there has been considerable interest in using this effect in metal base transistors (Sze 1969). To date, the more important properties of the Schottky barrier are its speed of response arising from unipolar current flow and low forward voltage drop compared with a p-n junction (Rhoderick 1978).

The Schottky barrier is not the only unipolar diode capable of injecting hot electrons. Current transport in certain types of t :terojunctions (Allyn et al 1980, Heiblum 1981) and bulk unipolar diodes is mostly due to the transport of one carrier type and provided the electric fields in the barrier region are high one must expect hot electron injection into the semiconductor. The bulk unipolar camel diode, for example, (Fig. 1b) attempts to maximise the quality of the diode by maximising the number of ionised impurities and the electric field in the barrier region and therefore embodies the main features required to form a hot electron emitter in a semiconductor.

An efficient hot electron emitter must be capable of injecting hot electrons into the neutral semiconductor adjoining the barrier with an energy $E_O = q\phi_B$ where $\phi_B$ is the height of the potential barrier. For a camel diode this will occur provided the width of the space charge region on the high field side of the barrier is less than the hot electron mean free path L.

$$\left( \frac{2\,\epsilon\,\epsilon_O\,\phi_B}{q\,N_A} \right)^{\frac{1}{2}} < L \qquad \qquad \ldots\ldots (1)$$

where $N_A$ is the acceptor concentration in the barrier region (Fig. 1). Taking L to be equal to 75 $\overset{\circ}{A}$ and a barrier height of 1 eV one obtains $N_A > 2 \times 10^{19}$ cm$^{-3}$ which should easily be achievable in silicon but is marginal in GaAs.

A further desirable requirement for an efficient hot electron emitter is that the current density of hot electrons is a maximum for a given barrier height and forward bias on the diode. As in a Schottky diode this occurs when current transport over the barrier is limited only by the velocity of the electrons and conditions around the top of the barrier satisfy the conditions for thermionic emission. Using Bethe's criterion (Bethe 1942) that the potential around the top of the barrier should change by kT within a mean free path, the doping concentration in the barrier region must satisfy the condition,

$$2\left( \frac{2\,\epsilon\,\epsilon_O\,kT}{q^2\,N_A} \right)^{\frac{1}{2}} < \lambda \qquad \qquad \ldots\ldots (2)$$

where $\lambda$ is the mean free path for warm electrons at the top of the barrier. The calculations shown in Fig. 2 indicate that conditions where thermionic emission rather than diffusion dominates the transport process occur when $N_A > 10^{18}$ cm$^{-3}$ but there is no range of $N_A$ where equation 2 is easily satisfied at room temperature. Computer calculations using the fundamental program POWTHY (Engl 1977) confirm that at least for the specific diode profile shown in Fig. 3, current transport is mostly determined by thermionic emission. In this example the quasi-Fermi level for electrons remains almost flat under a forward bias of 150 mV indicating that the concentration of electrons at the top of the barrier changes exponentially with bias. The doping concentration at the top of the barrier is $\sim 10^{18}$ cm$^{-3}$ which satisfies equation 2.

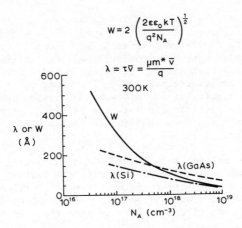

$$W = 2 \left( \frac{2\varepsilon\varepsilon_0 kT}{q^2 N_A} \right)^{\frac{1}{2}}$$

$$\lambda = \tau \bar{v} = \frac{\mu m^* \bar{v}}{q}$$

Fig. 2    Electron mean free path $\lambda$ and width of the barrier around the potential maximum as a function of the acceptor concentration in the barrier region of a bulk unipolar camel diode.

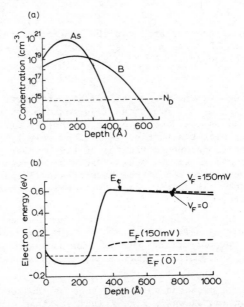

Fig. 3    Computed band shape and position of quasi-Fermi level for electrons under 150 mV forward bias for the impurity profiles shown in (a).

Compared with a Schottky diode, the bulk unipolar diode has, in general, a poorer blocking characteristic because the barrier is more easily 'pulled down' with reverse bias. However, as shown in Fig. 4 the softer camel barrier can in fact have harder blocking characteristics with low substrate fields because the image force has a negligible effect on the barrier height compared with its effect on the Schottky barrier. The calculations shown in Fig. 4 assume that minority carriers do not accumulate in sufficient numbers in the potential minimum to affect the

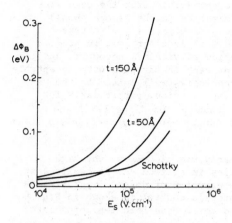

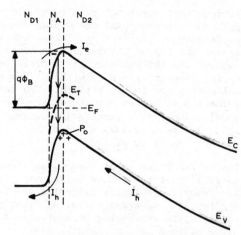

Fig. 4   Change in barrier height
ΔV against maximum electric field
in the lightly doped side of the
barrier.  t is the thickness of
the fully depleted $p^+$ layer.

Fig. 5   Schematic showing hole
accumulation at the potential
minimum under reverse bias
$N_{D1} \gg N_{d2}$.  $E_T$ is a
recombination level.

barrier height.  This is clearly not always the case particularly when
large numbers of holes are generated optically (Chen et al 1981,
Georgoulas 1982) or  via forward biassed p–n junctions (Board et al 1981)
and barriers are used as optical detectors or switching devices.

Assuming that the only mechanism for removing holes is by emission out of
the potential minimum (Fig. 5) a condition for negligible hole storage and
insignificant modulation of the barrier height is,

$$N_A \gg P_o \approx \frac{\sqrt{3}\,I_h}{4\,I_e}\,N_C\left(\frac{\tau_L}{\tau_{COLL}}\right)^{\frac{1}{2}} \qquad \ldots\ldots (3)$$

where $I_h$ is the hole current and $I_e$ is the electron current passing over
the barrier (thermionic emission is assumed for the electrons), $N_C$ is the
density of states in the conduction band, $\tau_L$ the lifetime of holes in
the degenerate n–type region, and $\tau_{COLL}$ is the mean time between
collisions.

Equation 3 shows that for typical leakage currents in well processed
devices the effect of hole storage is negligible for low barriers but
becomes increasingly important as the barrier height increases.  With a
generation current of $10^{-8}$ amps $cm^{-2}$, for example, equation 3 fails
for $\phi_B > 0.6$ eV.  In practice this is pessimistic because holes will
also be removed due to recombination with electrons via deep levels in the
highly doped barrier region (Fig. 5).

## 3. Fabrication of Bulk Unipolar Structures – Nanometre Engineering

The fabrication of bulk unipolar hot electron structures requires the
formation of narrow, highly doped layers in a semiconductor with abrupt
transitions from one doping level to another.  From the above, it is

evident that layer thicknesses should be comparable with the mean free path for electrons and the doping concentration in the layers should preferably be $\approx 10^{19}$ cm$^{-3}$. Taking the mean free path for electrons around the top of the barrier to be $\approx 100$ Å it is clear that we need to control the thickness of the barrier region to within a nanometre and the doping concentrations to within a few per cent if barriers with acceptable uniformity are to be obtained. Two technologies well suited to the formation of narrow structures are ion implantation, particularly for precise impurity concentration control (Shannon and Clegg 1983) and molecular beam epitaxy (MBE) because it can provide very precise thickness control (Neave et al 1983).

Ion implantation in silicon is particularly useful because the high solubility of the common dopant impurities in silicon can be realised at low temperatures where diffusion and interaction of dopant impurities is negligible. An example showing tight profile control is shown in Fig. 6 where layers containing both arsenic and boron have been implanted into <100> silicon at low energies and annealed at 600°C. Solid phase regrowth of the layers occurs at temperatures above 500°C to give high electrical activities well above solid solubility in some cases with no measurable movement or interaction of the impurities to within the resolution of the SIMS measurement ($\approx 3$ nm). This very encouraging result shows that it should be possible to form multilayer structures using solid phase regrowth of amorphous implanted structures but because of the increase in straggle of the implanted atoms with mean range it is not possible to fabricate more than 3 or 4 layers using direct implantation alone (Fig. 7). However, work on the recrystallisation of amorphous silicon on silicon (Hung et al 1980, Ishiwara et al 1982) and epitaxial growth of silicon on silicon at low temperatures (de Jong et al 1983, Milosavljevic et al 1983) shows that it is possible to fabricate structures using a combination of ion implantation and epitaxial growth. One reason for this result is the extremely good stability of the silicon surface. TEM and lattice imaging of implanted layers show that the layers recrystallise to give single

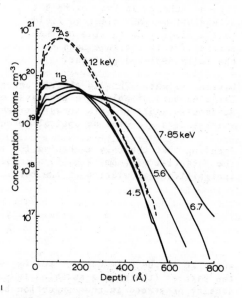

Fig. 6   Concentration profiles obtained using SIMS measurements on samples pre-amorphised with 30 keV Si. The two dashed curves are arsenic profiles before and after solid phase regrowth at 600°C. The two full curves at 4.5 keV are the equivalent for boron. The remaining boron profiles were all measured after annealing at 600°C (Shannon and Clegg 1983).

crystal layers with no extended defects up to the first atomic layers at the surface (Fig. 8) despite the presence in some cases of several monolayers of carbon and oxygen on the silicon surface. A further potentially important property of implantation into silicon is the ability

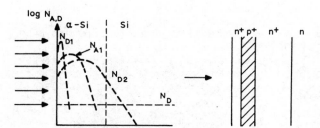

Fig. 7 Schematic diagram illustrating formation of nanometre structures using solid phase regrowth. $N_D$ is the background doping concentration.

to form highly doped metastable layers with impurity concentrations well above equilibrium solubility. This property can be easily incorporated when using heat pulse or rapid electron beam annealing (McMillan et al 1983) because diffusion and precipitation of impurities are prevented.

An example of the forward I/V curves measured on a silicon bulk unipolar camel diode formed using solid phase regrowth of an implanted structure is shown in Fig. 9. The annealing temperature was marginally higher than that required to prevent diffusion and the acceptor concentration at the potential maximum was lower than required to make a high quality hot-electron diode. Nevertheless, ideality factors were $\approx 1.1$ and extended over several decades of current at low temperatures where conditions for thermionic emission are more likely to apply. A further important feature of these implanted diodes was the ability, in some cases, to control the spread of barrier heights over a sample to within $\approx kT$ at room temperature.

At the present time MBE is better suited to the formation of hot electron structures in the III-V compounds than the technique of ion implantation. In general, activation of implanted impurities is more difficult to achieve and temperatures are required which produce some degree of dissociation and defect formation at the surface of the material no matter what capping layer is used. In contrast, doping via effusion cells at low temperatures during epitaxial growth has been shown to produce narrow, highly doped layers with abrupt transitions between doping levels and excellent thickness control. Several types of bulk unipolar diode have been made in GaAs including planar doped barriers, p-plane barriers and camel diodes. The work on camel diodes (Woodcock and Harris 1983) shows that diodes of high quality can be made using MBE (Fig. 10) over a range of barrier heights. It is interesting to note that even with the high barriers shown in Fig. 10 there is no evidence that the leakage current modulates the barrier height.

## 4. Gate Controlled Hot Carrier Devices

Since a bulk unipolar diode is formed by a narrow plane of ionised impurities within a semiconductor, it should be possible to control the barrier electrostatically via a gate located in the vicinity of the barrier region. Kazarinov and Luryi (1982), for example, have proposed a gate controlled thermionic emission transistor in which a gate on the surface of a semiconductor influences the barrier height of a planar

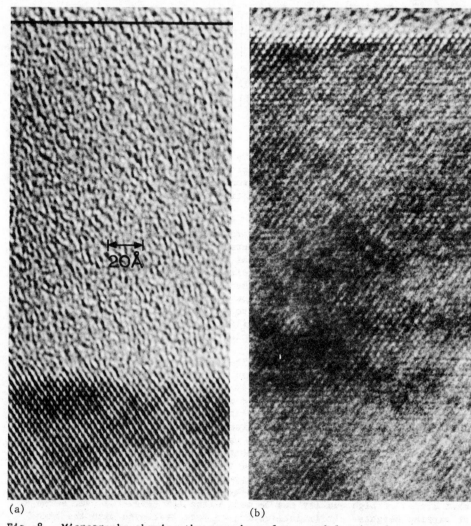

(a)                                          (b)

Fig. 8    Micrographs showing the amorphous layer and lattice fringes in silicon   (a) after bombardment with $2 \times 10^{15}$ As atoms $cm^{-2}$ at 10 keV (b) after annealing the implanted layer at 900°C using a multiscan electron beam with a dwell time of 100 ms.   The sheet resistance of the layer was $2 \times 10^{-4}$ $\Omega cm$ (McMillan 1983).

barrier located beneath it.   The structure is shown schematically in Fig. 11 together with a gate controlled hot carrier transistor where a bulk unipolar diode is controlled using capacitive coupling via an MOS gate laterally spaced from the barrier at the surface of the semiconductor. Compared with an MOS transistor these structures have the potential advantage of a higher mutual conductance $g_m$ because the source-drain current changes exponentially with barrier height at high current

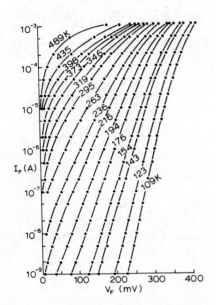

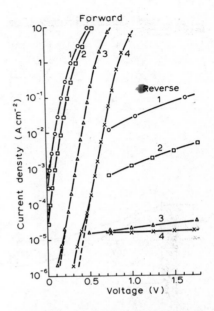

Fig. 9   Forward characteristic of a bulk unipolar diode in silicon formed by implanting boron at 4.5 keV and arsenic at 10 keV and annealing at 750°C (Shannon and Goldsmith 1983).

Fig. 10   Forward and reverse characteristics of bulk unipolar camel diodes in GaAs made using MBE. The thickness of the p-layer varied from 140 Å ( 1) $\phi_B$=0.55 V to 230 Å ( 4) $\phi_B$=0.94 V $N_A$ was $\approx 3\times10^{18}$ cm$^{-3}$ (Woodcock and Harris 1983).

densities where the MOS transistor is strongly inverted. This feature is clearly shown in Fig. 12 where the calculated current between anode and cathode for a gate controlled hot carrier transistor is compared with the current through an MOS transistor having the same oxide thickness (150 Å). It is seen that while the current in an MOS transistor increases exponentially with gate voltage in the sub-threshold region, the onset of inversion leads to the familiar quadratic dependence of current with gate voltage and an associated $g_m$ which is poor in comparison with the bipolar transistor. The gate controlled hot carrier device is able to maintain exponential behaviour up to higher current densities and the $g_m$ increased to $\approx$400 mS/mm compared to 100 mS/mm for the MOST. Gate control of camel diode structures has been demonstrated on the first experimental structures (Brotherton et al 1983) (Fig. 13) but considerable improvement in the technology is required before any of the predicted advantages can be realised.

## 5. Monolithic Hot Electron Transistors

The ability to fabricate hot electron diodes in the bulk of a semiconductor offers the possibility of realising monolithic hot electron transistors in which hot electron emission, transport and collection takes

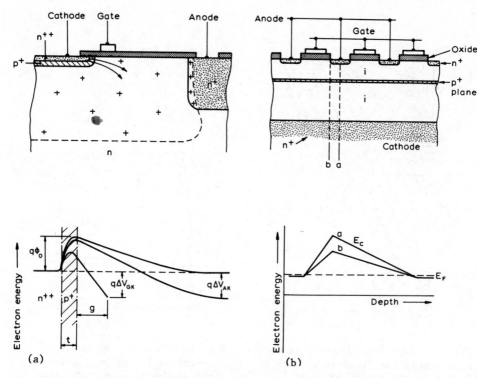

Fig. 11    a) A gate controlled hot electron transistor
            b) A gate controlled thermionic emission transistor.

place in the same semiconductor material (Shannon 1979b). Many structures
may be envisaged but basically they all consist of a bulk unipolar diode
for emitting hot electrons, a narrow degenerate base region and a bulk
unipolar collector diode (Fig. 14). The advantage of these structures
compared with metal base, hot electron transistors investigated in the
past (Sze 1969, Sze and Gummel 1966) derives from the use of a single
semiconductor material with continuity in lattice structure. This
property is particularly important for the collector barrier because it
must be a very efficient collector of hot electrons if useful current
gains are to be achieved. The excellent quantum mechanical transmission
coefficient of a bulk unipolar diode, even when assuming a square barrier
is shown in Fig. 15. In comparison, the transmission of a Schottky
barrier is very poor which is one of the main reasons for the demise of
the metal base transistor.

The most important property of the hot electron transistor is its ability,
in principle, to operate with a high frequency performance superior to a
bipolar transistor. This derives from minimal emitter and collector
capacitances and unipolar current transport up to high current densities
together with a mutual conductance which, just like a bipolar, increases
exponentially with emitter current. With these features it is easily
calculated (Shannon 1981) that provided a large enough base transport
factor can be achieved, optimum transistors should operate in the
50 - 100 GHz region for both silicon and GaAs.

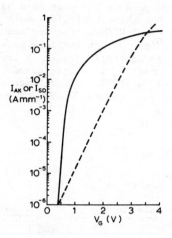

Fig. 12   Comparison between the source-drain current of a MOSFET (W = 0.5 μm to $t_{ox}$ = 150 Å) and the anode-cathode current of a gate controlled H.E.T. $t_{ox}$ = 150 Å.   The thickness of the $p^+$ layer was asssumed to be 150 Å.

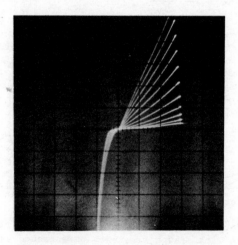

Fig. 13   Characteristics measured on an experimental gate controlled hot electron transistor.   The thickness of the $p^+$ barrier region was ≈300 Å, $t_{ox}$ was 300 Å.   x=1V/div, y=0.1mA/div, 2V/step on gate.   (Brotherton et al 1983).

The two structures shown schematically in Fig. 14 are designed to use two different features of hot electron transport, namely ballistic transport and diffusion. The width of the base needs to be small compared with the momentum relaxation mean free path in the case of a ballistic transistor or small in comparison with the characteristic length for energy loss in the case of diffusive transport. Clearly for optimum performance from a ballistic transistor one requires a high mobility material and low barriers so as to minimise the scattering rate. A transport factor of 0.74 has been measured, for example, in planar doped GaAs transistors with barrier heights <0.35 eV (Hollis 1983). Measurements made at 77 K on these transistors indicated that $F_T$ was ≈50 GHz for current densities >$10^4$ amps cm$^{-2}$.

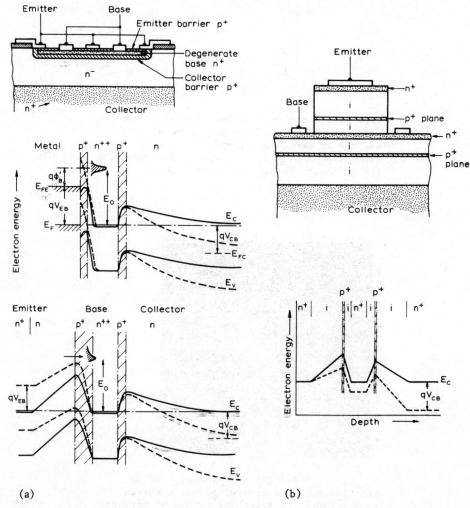

Fig. 14   Structures and band diagrams for  (a) monolithic hot electron transistors based on bulk unipolar camel diodes with tunnel emitter and camel emitter  (b) ballistic transistors based on planar doped barriers (Malik 1981).

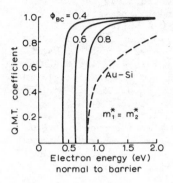

Fig. 15    Quantum mechanical
transmission across a bulk unipolar
diode with an abrupt potential barrier
and zero field.  A metal semiconductor
barrier is shown for comparison
($m_1^* = m_2^*$) (Shannon and
Goldsmith 1982).

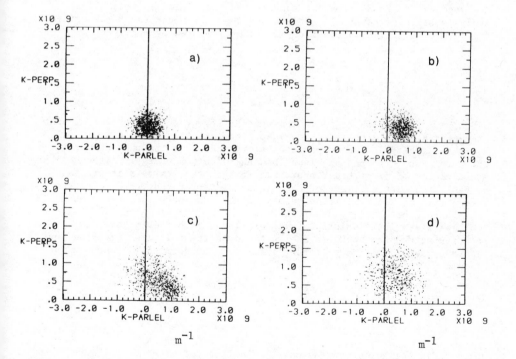

Fig. 16    Response of electrons in silicon to an electric field step from
0 to $4 \times 10^4$ V cm$^{-1}$  (a) t=0  (b) t=0.1 ps  (c) t=0.2 ps  (d) t=1.0 ps.
K-PARLEL is K vector in the x direction parallel to the electric field.
K-PERP = $(K_y^2 + K_z^2)^{\frac{1}{2}}$ (Yorston 1983).

In a low mobility semiconductor, such as silicon, one cannot consider a purely ballistic approach because the momentum relaxation mean free path is too short even at low energies. One therefore has to rely on hot electron diffusion from an emitter barrier. Fortunately because the solubility of impurities in silicon is high and low resistivities can be obtained, the based width can be reduced to 100 Å while preserving a base resistance of a few hundred ohms per square. In a silicon monolithic hot electron transistor, therefore, electrons pass over an emitter barrier, are accelerated by the strong electric field before entering the neutral base at an energy equal to the emitter barrier and a velocity well in excess of the steady state value. Velocity overshoot in silicon (Constant 1980) is well pronounced even for fields as low as $4 \times 10$ V.cm$^{-1}$ (Fig. 16) where the mean velocity in the direction of the field is $2 \times 10^7$ cm.sec$^{-1}$. Electrons then lose energy due to the generation of optical phonons at a rate dependent upon their excess energy above the Fermi level but the base width is sufficiently thin for most of the electrons to pass over the collector barrier before thermalising.

An example of the impurity profiles in a silicon structure formed using ion implantation is shown in Fig. 17. The active base width is 200 Å and is bounded in this case by a 1.0 eV emitter barrier and a 0.5 eV collector barrier. The emitter uses a reverse biassed metal-semiconductor junction through which electrons tunnel via field-emission (see Fig. 14).

The diffusion of hot electrons across a semiconductor base has been analysed using the Boltzmann transport equation (Ridley 1981). The analysis assuming parabolic bands shows that transport can be described in the form of a diffusion equation with time replaced by energy

$$\frac{\partial \psi}{\partial E} (E, \ x) \ + \ \frac{L^2}{\hbar \omega} \ \frac{\partial^2 \psi}{\partial x^2} (E, \ x) \ = \ 0 \qquad \ldots \ldots (4)$$

$\psi(E,x) = E \ f_o(E,x)$ where $f_o$ is the probability of occupancy of each state at energy E, L is a hot electron mean free path (corrected for non-parabolicity in the comparison below) and $\hbar \omega$ is the phonon energy which is assumed to be much less than the electron energy. Although there are a number of first order approximations in the analysis it provides probably the best estimate of the transport factor that we have at the present time. The calculated values of the transport factor are shown in Fig. 18 for the boundary condition $\psi = 0$ at $x = \infty$ (a) and $\psi = 0$ at $x = W$ the base width (b). Solution (b) (Berz 1983) is probably more representative when there is a high field in the collector sweeping the electrons away from the collector barrier. The normalised length z is given by

$$z \ = \ W \left/ \left[ 2L \left[ \left( E_o - q \ \phi_{BC} \right) \middle/ \hbar \omega \right]^{\frac{1}{2}} \right] \right. \qquad \ldots \ldots (5)$$

where $E_o$ is the height of the emitter barrier and $q\phi_{BC}$ the collector barrier.

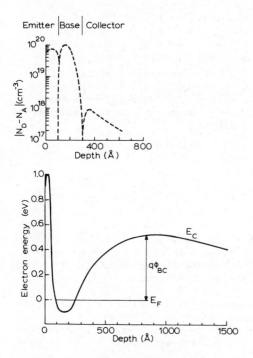

Fig. 17    Impurity profiles in a
silicon hot electron transistor
measured using SIMS and the
corresponding potential profile.

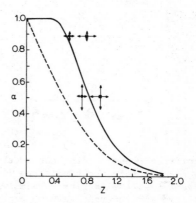

Fig. 18    Calculated base
transport factor $\alpha$ against a
normalised length Z (Ridley 1981)
see equation 5.  The experimental
points are for silicon
transistors made using ion
implantation ●, w = active base
width ■, w = base width to 50% of
collector barrier height.

The agreement between the calculated and measured values is reasonable in
view of the approximations used;   clearly the calculations underestimate
the transport factor even assuming the more favourable boundary
condition.  The sheet resistance of the base was $\approx 600$ $\Omega/\square$.  The best of
the measured results are very encouraging because they show that an $\alpha$ of
0.9 can be obtained in structures having a low intrinsic base resistance.

The measured values of $F_T$ on the better devices made so far are plotted
against current density in Fig. 19 and compared with the calculated
performance of optimised structures (Shannon 1981).  Clearly the use of a
tunnel emitter severely degrades the high frequency performance because of
the very high emitter capacitance and measured values are close to those
calculated.  The use of a bulk unipolar camel emitter should increase $F_T$
very considerably due to its much smaller capacitance (Shannon and Slatter
1983).

6. Conclusions

It is concluded that it should be possible to fabricate hot-electron
diodes in silicon and GaAs capable of efficient injection of hot electrons
into an adjoining semiconductor region.  Silicon is very promising within
this context because the high solubility of the common dopants enables
high electric fields to be engineered within the semiconductor and

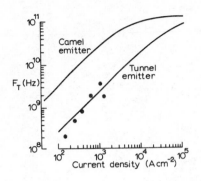

Fig. 19    Measured $F_T$ against current density for a monolithic hot electron transistor with a tunnel emitter compared with the calculated performance.

significant band bending can occur over extremely small distances.  Diodes may also satisfy the conditions for thermionic emission in which case current over the barrier is limited only by the thermal velocity of the carriers and the voltage drop across the diode is minimal.

Both ion implantation and molecular beam epitaxy are suitable technologies for fabricating the nanometre structures required; the former is particularly suitable for making structures in silicon while the latter is well suited to device fabrication in the III–V compounds.

Several transistor structures have been proposed which make use of hot electron diodes either as fast unipolar devices through which the current can be controlled using a gate or as emitters and collectors of hot electrons inside the material.   The predicted performance of optimised structures is considerably in excess of that achievable with conventional MOS and bipolar transistors.

Encouraging results have been obtained in both GaAs and silicon.  Hot electron diodes have been made in both materials with ideality factors of 1.1 and with good blocking characteristics and the spread of barrier heights has, in some cases, been comparable with kT at room temperature.

Gate controlled hot carrier devices have given transistor action and monolithic hot electron transistors have been made in silicon and GaAs with current gains and high frequency performances superior to any hot electron transistor made in the past.

Acknowledgements

It is a pleasure to acknowledge many colleagues at PRL for their contribution to this work.  In particular I am indebted to J.A.G. Slatter for modelling the camel diode using the computer program POWTHY and Dr F. Berz for obtaining solutions to equation 4.  I am also grateful to Dr S.D. Brotherton of PRL, R. Yorston of University of Reading, G.B. McMillan of University of Cambridge and Dr M. Hollis of Cornell University for providing unpublished results.

Part of this work was carried out under contract to the Procurement Executive, Ministry of Defence, D.C.V.D., London.

## References

Allyn C L, Gossard A C, Wiegmann W 1980 Appl. Phys. Lett. <u>36</u> p373.

Berz F 1983 Private Communication.

Bethe H A 1942 M.I.T. Radiation Lab. Report 43-12.

Board K 1982 Microelectronics Jr. <u>13</u> 3 pp19-22.

Board K, Darwish M 1981 Electronics Lett. <u>17</u> p41.

Brotherton S D, Gill A 1983 Private Communication.

Carroll J E 1970 Hot Electron Microwave Generators, Edward Arnold Ltd., London.

Carroll J E 1974 Physical Models for Semiconductor Devices, Edward Arnold Ltd.

Chen C Y, Cho A Y, Garbinski P A, Bethea C G, Levine B F 1981 Appl. Phys. Lett. <u>39</u> p340.

Constant E 1980 Inst. Phys. Conf. Ser. 57, pp141-168.

Engl W 1977 POWTHY was developed at Institute fur Theoretische Elektronik, Technische Hochschule Aachen.

Georgoulas N 1982 IEEE Electron Devices Lett. EDL-3, p.61.

Heiblum M 1981 Solid State Electronics <u>24</u> pp343-366.

Hollis M A 1983 Cornell University - Private Communication.

Hung L S, Lau S S, Von Allmen M, Mayer J W, Ullrich B M, Baker J E, Williams P, Tseng W E 1980 Appl. Phys. Lett. <u>37</u> p909.

Ishiwara S, Naruke K, Furukawa S 1982 Jap. Jr. Appl. Phys. Letter <u>9</u> pL577.

de Jong T, Douma W A S, Doorn S, Saris, F W 1983 Materials Lett. 1 p157.

Kazarinov R F, Luryi S 1982 Appl. Phys. <u>A28</u> pp151-160.

Mader H 1982 IEEE Trans. <u>Ed-19</u> 11 p1766.

Malik R J, AuCoin T R, Ross R L, Board K, Wood C E C, Eastman L F 1980 Electronics Lett. 16 pp836-837.

Malik R J, Board K, Eastman L F, Wood C E C, AuCoin T R, Ross R L 1981 Inst. Phys. Conf. Ser. Vol. 45, p697.

Malik R J 1981 Gallium Arsenide Planar Doped Barrier Diodes and Transistors grown by MBE - Thesis Cornell University.

McMillan G B 1983 Cambridge University Engineering Dept., Private Communication.

McMillan G B, Shannon J M, Ahmed H 1983 Materials Research Soc. Symposium Proceedings Vol.13 pp437-442.

Milosavljevic M, Jeyens C, Wilson I H  To be published.

Neave J H, Joyce B A, Dobson P J, Norton N 1983 Appl. Phys. $\underline{A31}$ pp1-8.

Rhoderick E H 1978 Metal-Semiconductor Contacts, Clarendon Press, Oxford.

Ridley B K 1977 Jr. Appl. Phys. $\underline{48}$ pp754-764.

Ridley B K 1981 Solid State Electronics $\underline{24}$ pp147-154.

Shannon J M 1979a Appl. Phys. Lett. $\underline{35}$ p63.

Shannon J M 1979b IEE J. Solid State Electron Devices $\underline{3}$ p142.

Shannon J M 1981 IEE Proc. $\underline{128}$ pp134-140.

Shannon J M, Clegg J B 1983 Nanometre Structures in Semiconductors Formed by Low Energy Ion Implantation.  To be published in VACUUM.

Shannon J M, Goldsmith B J 1982 Thin Solid Films $\underline{89}$ pp21-26.

Shannon J M, Slatter J A G 1983 Jap. Jr. Appl. Phys. Sup. 22-1 p259.

Shannon J M, Goldsmith B J 1983  To be published.

Sze S M, Gummel H K 1966 Solid State Electronics $\underline{9}$ pp751-769.

Sze S M 1969 Physics of Semiconductor Devices, Wiley & Son.

Woodcock J M, Harris J J 1983 Electronics Lett. $\underline{19}$ 5 p181-183.

Yorston R 1983 Computer Science Dept. University of Reading - Private Communication.

*Inst. Phys. Conf. Ser. No. 69*
*Paper presented at ESSDERC/SSSDT 1983, Canterbury 13–16 Sept. 1983*

63

# Hot carrier injection in oxides and the effect on MOSFET reliability

P.Balk

Institute of Semiconductor Electronics / SFB 202, Aachen Technical University, D-5100 Aachen, FRG

Abstract. Capture of hot electrons and holes injected into the gate oxide of short-channel MOSFETs leads to degradation of the device characteristics. This review focusses on one aspect of this problem, namely the effect of the technology used in preparing the gate system on the trapping properties of the insulator. The main topics treated are the effects of dopants, irradiation, high temperature annealing and of the injected carriers themselves on the bulk and interfacial trapping behavior and on the dielectric strength of the gate system. The requirements for an optimized gate technology are discussed.

## 1. Introduction

Exact control of device threshold voltage and transconductance are essential requirements for the design of optimized silicon MOSFET integrated circuits. Thus, it is not surprising that in the history of the development of the MOS technology problems relating to the characteristics of the gate system have played a dominant role. These problems were generally related to interface states, insulator charges and the dielectric properties (polarizability, dielectric strength) of the insulator system.

In the sixties the problem of the control of interface states and fixed oxide charge centers was mastered by the application of suitable annealing treatments. The dangling silicon bonds at the interface were chemically saturated by forming Si-H bonds (for example in a post-metallisation anneal in the Al gate technology). A high temperature anneal in neutral ambients after oxidation effectively removed all fixed insulator charges and supposedly also the corresponding defect centers. The use of clean processing conditions and/or the application of a thin phosphosilicate glass layer solved the mobile $Na^+$ ion problem and at the same time greatly improved the oxide quality with regards to low field breakdown and wear-out. For very high negative gate voltages "negative bias instability" would occur. This phenomenon, which was not understood at that time, leads to the build-up of positive charge in the oxide. However, the required fields were quite high ($> 5MV$ $cm^{-1}$) so that this effect could only play a role in the operation of non-volatile memory devices but would not be a major concern in standard MOSFETs. It was thought that charge transfer only took place between the silicon and immediately adjoining oxide centers, for example by a hopping process (Hofstein 1967).

The developments of the seventies towards short-channel devices for higher packing densities and larger circuit speeds gave rise to a new class of problems. At channel lengths below approx. 2μm the limits of the linear scaling regime, in which field values are not affected by miniaturization of the devices, are reached. Due to superlinear increases in channel doping and non-scaling operating voltages short-channel MOSFETs exhibit increased fields parallel to the Si-SiO$_2$ interface and consequently show carrier heating and multiplication effects in the channel, particularly near the drain. The drift of such hot carriers into the oxide leads to instability of the threshold voltage of the type reported by Hara et al.(1970) due to electron or hole capture and to the generation of interface states.

These instabilities are the topic of the present review. We will first discuss the different modes of charge carrier injection in the gate insulator of short-channel devices. In a following section we will review those features of the technology that are likely to create electron and hole trapping centers. We will show that those technology steps that increase the density of oxide traps also lead to enhanced sensitivity towards creation of interface states, decrease the radiation hardness of the MOS structure and promote the incidence of low field breakdown. The optimization of the technology with regards to charge carrier capture at room temperature and at liquid N$_2$ temperature will be discussed. This review will finally deal with the question of the further development of the MOS technology in view of the problems discussed in this paper.

## 2. Hot Carrier Injection and MOSFET Degradation

Degradation of MOSFET characteristics is mainly observed when carriers enter into the bands of the SiO$_2$. The latter process most easily takes place for carriers with sufficient energy to cross the electron or hole barriers at the surface. In this case only a small component of electric field perpendicular to the Si-SiO$_2$ interface is necessary for injection to occur. Such injection currents have indeed been measured directly as gate currents; it is also possible to determine the charge trapped in the gate insulator or the charge collected on the gate electrode for the case where the electrode is kept floating.

Four injection modes have been reported for n-channel MOSFETs (fig. 1): The first mode (a) is the injection of electrons heated up in the channel, particularly near the drain upon onset of pinch off. The second injection mode (b) is that of electrons thermally generated in the substrate near the surface and accelerated towards the Si-SiO$_2$ interface for large values of the substrate bias voltage (Ning et al. 1977, Ning et al. 1979, Cotrell et al. 1979). For conditions where the channel near the drain is heavily avalanching (c) both electron and hole injection may occur (Cottrell et al. 1979, Takeda et al. 1983). A final mode (d) occurs when the substrate hole current produced by avalanching near the drain generates further electron-hole pairs and these secondary electrons are injected into the oxide as in the case of substrate hot electron injection. This type of injection becomes particularly pronounced for large substrate bias voltages (Takeda et al. 1983). Carrier injection in the oxide of field effect transistors is a localized phenomenon; it takes place over only a fraction of the total length of the channel.

Because of the direction of the field in the oxide hole injection can only take place close to the drain; injection of electrons requires the opposite field and will be more pronounced in the central region of the channel. In principle similar phenomena will occur in p-channel devices with

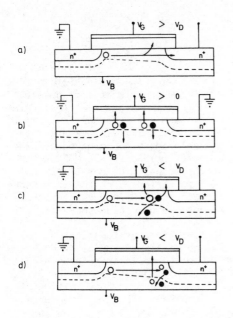

Fig. 1   Electron (o) and hole (●) injection in the oxide of n-channel MOSFETs

the roles of holes and electrons exchanged. However, exact symmetry in the behavior of n and p channel devices can not be expected and is indeed not observed (Ng and Taylor 1982).

One reason is the different barrier heights for electrons and holes at the Si-SiO$_2$ interface (3.1 and 4.7 eV, respectively). A further reason is the lower effectiveness for pair production of holes in the channel (Kamata et al 1976). Also, the interaction of the two types of carriers with SiO$_2$ is rather different, the holes leaving more "damage" upon capture, for example in the form of interface states.

Generally, the injection of carriers causes a combination of transconductance degradation and threshold shift. The classical interpretation has been that in n-channel devices these effects were due to electrons (Ning et al. 1977). In recent years a discussion on the injection of holes in n-channel devices has started. Charging effects and transconductance changes appear to support the idea that even though hole gate currents may be too small to be measured directly in most cases their effect on device properties may be considerable (Fair and Sun 1981, Gesch et al.1982, Hofmann 1983, Borchert et al.1983, Bauer and Balk 1983).

The device degradation discussed in this section results from a combination of two factors: The generation of hot carriers and their capture in the oxide. Since avalanche breakdown at the drain junction plays an important role in the generation process a number of attempts have been made to suppress this effect. One rather effective solution is to use a more gradual junction by applying a double diffusion process (As+P) or by using P only (Ogura et al.1980, Takeda et al.1982). Improved stability is also obtained by decreasing the effect of the gate on the field at the drain by using a burried channel (Mizutani et al.1981; Takeda et al.1982) or by offsetting the gate (Takeda et al.1982). However, the technology of the channel region offers only limited room for optimization of device properties.

The charge trapping behavior of the oxide is the second factor to consider. Also in this case the requirements of the total device fabrication process limit the possibilities for optimization of the oxide characteristics. This aspect is the major topic in the remainder of this review.

In evaluating the properties of the MOS gate system most data have been obtained from measurements on capacitors or devices using dc stress conditions. However, it has been found that in some cases larger shifts in the device characteristics are obtained when the stress is applied in the form of pulses (Fair and Sun 1981, Bauer and Balk 1983). It thus should be kept in mind that dc stress measurements tend to offer a relatively optimistic

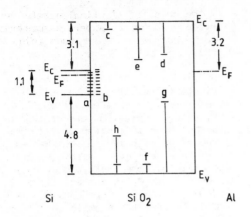

Fig. 2 Barriers and traps in MOS system;
a,b: fast, slow interface traps; c,d,e :
electron traps; f,g,h: hole traps

view of MOSFET stability.

## 3. Trapping Centers in the Si-SiO₂ System

The wide bandgap of $SiO_2$ offers ample opportunity for defect states with one or more levels to manifest themselves (Bennett and Roth 1971). They are always related to the presence of imperfections in the MOS system, be it point defects (intrinsic or extrinsic) or the interfaces between $SiO_2$ and the substrate or the gate electrode (fig. 2).

Interface states located energetically in the Si bandgap can change their state of charge, like other states in the Si gap, upon changing the position of the Fermi level at the surface by application of a voltage to the gate electrode. Depending on the rate at which this charge exchange takes place they are characterized as fast or slow. Most likely the difference in response time is related to the distance of the centers from the metallurgical interface, the slow states being located at least one atomic distance inside the oxide. The bulk states extend into the bulk of the film (not necessarily uniformly); they may capture electrons or holes. These centers do not exchange charge with the silicon, either because they are at least a tunneling length away from it or because their energy position is located outside the region of the Si bandgap.

Density and energy distribution of interface states are determined from the shape of the C-V curve. Most information on the bulk centers has been obtained by avalanche injection of known amounts of electrons or holes into the oxide and monitoring the state of charge of the system from the position of the C-V curve. However, C-V measurements provide only information about the so-called effective charge, which is the product of the total amount of captured charge in the oxide and the position of its centroid divided by the oxide thickness. To separate these factors additional measurements (of the photo I-V or dark I-V characteristics) are required. Analysis of the dependence of the experimentally observed effective charge on the density of injected carriers permits determination of the effective densities and of the capture cross sections of the centers. Information on the trap depth  may be obtained from thermal detrapping measurements. It is customary to distinguish between deep and shallow traps, the shallow traps being those that can only be filled at 77 or 100 K. For further details on properties and measurements of traps the reader is referred to reviews by DiMaria (1978) and DeKeersmaecker (1983).

## 4. Effects of Processing on the Electrical Properties of the Gate Insulator

It is well known that there is a direct relation between processing conditions and defect structure of solid materials. In many cases these defects function as trapping centers for electrons or holes. The following processing induced defects are of interest for the MOS technology: The dopants used as implants for source, drain and the channel as well as the poly-Si gate dopants (As, P, B) will enter into the oxide during implantation, oxidation and high temperature annealing.

Even gate systems prepared under very clean conditions will contain a certain amount of defects related to $H_2O$. Unless extreme care is taken to work under ultra dry conditions $H_2O$ is taken up by the $Si-SiO_2$ system and firmly bonded into its structure, particularly in the region near the interface (Revesz 1979).

The above discussion could suggest that in the absence of such impurities the insulator is free of trapping centers. In comparison with other insulators this is certainly the case (Balk 1974, 1975, 1983). Silicon dioxide has been characterized as the ideal insulator because the relatively weak dependence of the Si-O bond energy on the O-Si-O bond angle permits the combination of a flexible overall structure where long range order is absent with chemical saturation of the oxide system (Revesz 1973). However, thermodynamics requires the presence of a number of defects even in pure oxides. Experience shows that this number may be very low, but it is finite. In addition, the above mentioned distribution of bond angles implies a distribution of bond strengths and thus also the presence of weaker bonds. These may particularly be expected in the strained transition regions between $SiO_2$ and the substrate and between $SiO_2$ and the gate electrode. Their depth into the oxide will be very shallow for oxides grown and eventually annealed at 900°C or below, where viscous flow does not smear out the strain over a larger depth (Irene et al. 1982).

The above mentioned intrinsic defects may be expected to be electrically active as trapping centers for electrons and/or holes. It will also be clear that the densities of the defects will be dependent on the temperature treatment of the MOS system. Thus the electrical characteristics will be determined by its annealing history.

Finally, by exposing the system to high energy radiation (be it of particle or electromagnetic nature) certain bonds that are relatively weak will break up and local structural rearrangement will take place. Given the increased importance of e-beam, x-ray, ion beam and plasma methods in modern microelectronic technology, the defects created by using such treatments are a further significant group of oxide centers.

In the following we will discuss the effects of the above mentioned impurities, of high temperature treatments and strain gradients and of energetic radiation on the "bulk" properties of the oxide. We will also examine the questions what conditions favor the creation of interface states at the $Si-SiO_2$ transition.

## 5. Process Induced Impurity Traps

The most obvious impurities to become incorporated into the oxide are those used in doping source, drain, channel region and the poly-Si gate film (in the poly-Si or polycide gate technology), namely As, P and B. These elements may enter the $SiO_2$ by implantation or during the formation of the gate insulator by thermal oxidation or during subsequent high temperature annealing of the gate system. It appears that the way these elements are being taken up into the oxide structure depends on the conditions under which the incorporation takes place. This point may be illustrated by the trapping behavior of $As^+$ implanted $SiO_2$. Upon annealing in $N_2$ at 1000°C, necessary to restore the damaged oxide, DeKeersmaecker and DiMaria (1980b) observed deep electron traps with cross sections in the range $10^{-17}-10^{-15}cm^2$ and a total density up to that of the implanted ions, but often somewhat smaller. Additionally, deep hole traps of large cross sections and with the same centroid were found by the same authors (DeKeersmaecker and DiMaria 1980a). These findings indicate that $As^+$ implantation produces amphoteric trapping sites. A similar behavior was reported for $P^+$ implanted samples; only the cross sections are smaller in this case (DeKeersmaecker and DiMaria 1980a). The electron and hole trapping behavior of annealed $P^+$ implanted films has recently been confirmed (Marczewski and Strzalkowski 1982 resp. Smeltzer 1982).

The way in which $P^+$ and $As^+$ are being incorporated into the oxide has been a matter of some discussion in recent years. Simple chemical considerations would seem to indicate that these elements substitute for Si. However, it was concluded on the basis of model calculations that P and As occupy O sites (DeKeersmaecker et al.1978, Pantelides 1982). Recent experiments by Alexandrova and Young (1983) show that As is indeed incorporated on Si sites: Their $N_2$ annealed samples which contain the above mentioned dominant electron traps with $\sigma =(1.5-2)\times10^{-15}$ and $(2-3)\times10^{-16}cm^2$ exhibit a decrease in the concentration of these centers when $Si^+$ is implanted in addition to $As^+$. This behavior is indicative of competition between As and Si for Si sites. The authors suggest that for increasing amounts of implanted $Si^+$ more and more As is forced on interstitial sites, where it is electrically inactive. In contrast, samples with $O^+$ implanted after As or As implanted samples annealed at high temperature in $O_2$ have a different defect structure: in this case a dominant electron trap with $\sigma=1\times10^{-15}$ $cm^2$ develops upon subsequent $N_2$ annealing. The center with $\sigma =(1.5-2)\times10^{-15}cm^2$ is still present but appears at reduced densities, that with $\sigma= (2-3)\times10^{-16}cm^2$ has disappeared altogether.

These findings point to a fundamental difference between O-deficient films (i.e. those with $Si^+$ implant and annealed in $N_2$) and the films which were after $As^+$ implantation first brought in a fully oxidized state by means of $O^+$ implantation or $O_2$ high temperature treatment before annealing them in $N_2$. This difference also shows up in the diffusion behavior of implanted $As^+$: at the concentrations of interest for our present discussion ($10^{15}$ $cm^{-3}$ or below) the dopant diffuses distinctly slower in an O-deficient oxide than in a fully oxidized film (Singh et al.1983b). Using simple model considerations one would expect that the fully oxidized samples contain configurations like (a) and (b) in fig. 3; O-deficient samples would, in addition to (a) and (b) also contain configurations like (c)-(e) plus possibly interstitial As and Si atoms. However, correlating these configurations with the observed trapping centers does not appear feasible.

In contrast to the above mentioned behavior of $As^+$ and $P^+$ implanted films, $B^+$ implantation apparently does not give rise to electrically active centers in the oxide (DiMaria 1978). Implantation through rather than into the $SiO_2$ layer was reported to lead to an electron trapping center with $\sigma=10^{-14}-10^{-15}cm^2$ (Marczewski and Strzalkowski 1982). However, since the films were annealed at only 500°C this center is probably caused by incomplete removal of implantation damage.

a) $O_3\equiv As-O-As\equiv O_3$

b) $O_3\equiv As-O-Si\equiv O_3$

c) $O_3\equiv As-Si\equiv O_3$

d) $O_3\equiv As\cdot$

e) $O_3\equiv Si\cdot$

Fig. 3 Expected configurations in fully oxidized (a,b) and O-deficient (c-e) As doped $SiO_2$

The presence of water related defects in $SiO_2$ even if the films have been grown under nominally dry conditions is well established by now (Revesz 1979). It has also been known for many years that the oxides grown under "wet" conditions or exposed to water vapor after growth at very moderate temperatures ($\leqslant 200°C$) contain considerably larger concentrations of deep electron traps than films obtained by oxidation in dry $O_2$ (Nicollian et al. 1969, 1971, Gdula 1976, Young et al. 1979). Hartstein and Young (1981) succeeded in identifying the water related electron traps by combining results of infrared absorption measurements with those of an avalanche injection study.

By varying the "water" content of the film by indiffusion of $H_2O$ or annealing in dry $N_2$ at 1000°C it was found that a center with $\sigma = 1 \times 10^{-17}$ $cm^2$ is associated with SiOH groups probably occuring in pairs; a trap with $\sigma = 2 \times 10^{-18}$ $cm^2$ is associated with the presence of loosely bound $H_2O$ in the oxide. The behavior of the centers suggests a reversible hydrolysis reaction involving a regular (strained?) Si-O-Si configuration. The concentration of $H_2O$ is in the $10^{17}$-$10^{18} cm^{-3}$ range; even in carefully prepared dry oxides it is still approx. $2 \times 10^{17}$ $cm^{-3}$. In contrast, the infrared measurements show that the concentration of SiOH can be varied considerably (from $10^{19} cm^{-3}$ to below $5 \times 10^{17} cm^{-3}$ which was the limit of detection) via the $H_2O$ vapor pressure in the ambient. As a third absorption center SiH was detected (concentrations always around $10^{16} cm^{-3}$, independent of heat treatment); however, a corresponding electron trap was not found, possibly because of its low concentration compared to that of the other centers. The $10^{-18} cm^2$ center is roughly uniformly distributed across the oxide film (Young et al. 1979), the $10^{-17}$ $cm^2$ trap probably has a distribution skewed towards the $Si-SiO_2$ interface (Feigl et al. 1981).

Upon annealing at high temperature (for example 1000°C) in neutral ambients the concentration of the deep water related electron trap is strongly reduced (Young et al. 1979). At the same time the hole trap concentration increases (Aitken and Young, 1977). This negative correlation suggests that the water related centers in all likelyhood do not act as hole traps. These centers, which even for dry oxides have relatively large densities, disappear rather rapidly (in 1-2 h) at 1000°C in $N_2$ (fig. 4, Aslam et al. 1983). It may also be seen that the density of the $10^{-19}$ $cm^2$ deep electron trap, an as yet unidentified center, decreases at a somewhat slower rate and stays at a finite value. Apparently the range of cross sections extends still further downward: Centers of unidentified

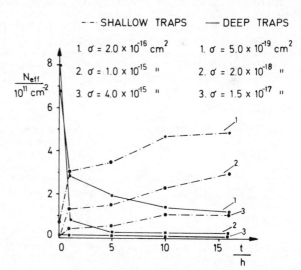

Fig. 4 Loss of deep $H_2O$ related traps and generation of shallow traps upon $N_2$ annealing at 1000°C

nature with values as low as $10^{-20}$ cm$^2$ have also been reported (Liang and Hu 1981, Sah et al. 1983). It was found that the $10^{-19}$ cm$^2$ center has a larger density in case of forming gas anneal, which will tend to reduce the oxide, than after a N$_2$ anneal (Young 1981). Finally, during prolonged high temperature N$_2$ annealing the uniform trap distribution changes to a U-shaped profile with a large built-up near the interfaces (Lai et al. 1981).

## 6. Generation of Traps by High Temperature Annealing

High temperature annealing in N$_2$ or Ar causes the development of shallow electron traps (fig. 4). Such shallow electron traps were first reported by Ning (1978) and by Aitken et al. (1978). In addition to the centers with $\sigma=10^{-15}-10^{-16}$ cm$^2$ shallow traps with $\sigma=10^{-19}$ cm$^2$ are also found (Singh et al. 1983a). The densities of the centers created by annealing increases with annealing temperature. It was also mentioned above that the density of the deep trap in the $10^{-19}$ cm$^2$ range is being reduced but does not disappear during this treatment; however, the distribution of the center across the oxide changes. A simple explanation would be that the deep center observed after annealing is a new one, possibly formed from the original centers. Since the rate of decay of the original centers and the appearance of the shallow traps exhibit the same time dependence the deep and shallow centers in the annealed samples may be structurally related. A correlation with the N or Na content of the films does not exist; thus it has been proposed (Aslam et al. 1983) that these defects are intrinsic in nature; they would be related to O-deficiency in the oxide and to chemical unsaturation caused by the split off of H$_2$O from neighbouring hydroxyl groups (fig. 5).

This concept fits the above observations. The stretched Si-Si bond was theoretically treated by Ngai and White (1981). These structures lead to levels in the gap of the SiO$_2$. Dangling oxygen bonds were invoked by Bhattacharyya et al. (1982) to explain electron trapping in oxides prepared in the presence of chlorine. Stretched Si-Si bonds and probably also the configuration obtained after removal of water will show a Jahn-Teller type internal charge transfer and exhibit dipole character. This could explain the relatively large capture cross sections of some of the shallow centers (close to $10^{-14}$ cm$^2$, which is considered the lower limit of the coulombic range). The increase in density with annealing temperature is caused by a shift to the right of the redox equilibrium SiO$_2$+Si$\rightleftharpoons$SiO (Lander and Morrison 1962).

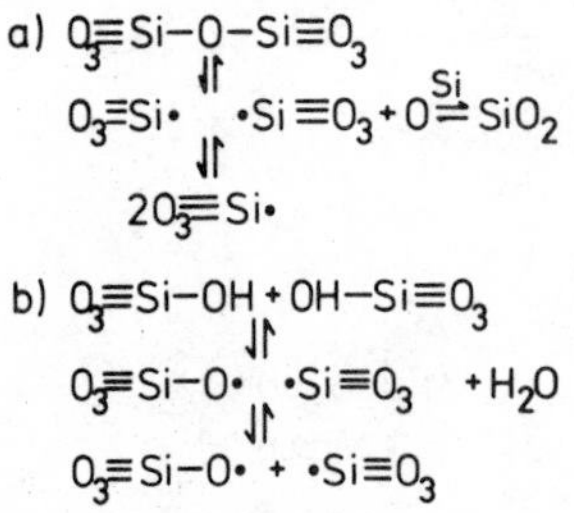

a) $O_3{\equiv}Si-O-Si{\equiv}O_3$

$O_3{\equiv}Si\cdot \quad \cdot Si{\equiv}O_3 + O{\overset{Si}{\equiv}}SiO_2$

$2O_3{\equiv}Si\cdot$

b) $O_3{\equiv}Si-OH + OH-Si{\equiv}O_3$

$O_3{\equiv}Si-O\cdot \quad \cdot Si{\equiv}O_3 \quad +H_2O$

$O_3{\equiv}Si-O\cdot + \cdot Si{\equiv}O_3$

Fig. 5 Loss of O (a) or H$_2$O(b) leading to stretched Si-Si and Si-O bonds and to dangling Si and O bonds

The model is supported by a number of further observations: Already at the temperature at which the shallow electron traps with $\sigma=10^{-16}$cm$^2$ are filled charge transfer to the $10^{-19}$cm$^2$ and smaller traps takes place (Aslam et al. 1983) indicating that the shallow and deep traps are levels of the same defect. Griscom and Fowler (1978) showed that such structures may indeed exhibit two levels within the Si bandgap. The small cross sections observed for the deep traps are probably caused by the fact that the deep levels are only accessible through shallow ones; the transfer may entail a relaxation of the bonding at the center and its surroundings. At room temperature and above the occupation probability of the shallow level is very low and consequently the cross section of the deep trap quite small.

In agreement with the above ideas a very low concentration of water rela- ted deep electron traps, but a relatively large concentration of the $10^{-19}$ cm² center along with the before mentioned shallow electron traps is found in poly-Si gate samples. The water related traps most likely disappear in the initial phase of the poly-Si deposition process. The densities of the- se intrinsic centers increase with annealing time with the poly-Si elec- trode in place. On the other hand, a brief $O_2$ anneal (typically 30 sec at 1000°C) before deposition of the metal in Al gate capacitors suffices to reduce the density of the centers by over an order of magnitude (Aslam et al. 1983).

The annealing treatments (Ar or $N_2$ post oxidation anneal; anneal with poly- Si gate) that lead to increased densities of intrinsic electron traps have been reported to also increase the density of deep hole traps (Aitken and Young 1977). Because of their large cross sections ($10^{-13}$-$10^{-14}$cm²) hole traps are very effective in capturing charge carriers. Comparing the to- tal density of shallow electron traps and that of the hole traps it is found that these are roughly equal. Both densities are reduced by the same brief $O_2$ treatment; however, in the case of the hole traps not quite to the same low level (Aslam and Balk 1983). Like the $10^{-19}$ cm² deep electron trap the hole traps show a large concentration near the Si-SiO₂ interface. The distribution of the shallow centers has not been determined. It was first suggested by Woods and Williams (1976) that the hole trapping centers are caused by excess Si. The similar annealing behavior indicates that these hole traps and the intrinsic electron traps are in all probability related to the same amphoteric defects. There are probably hole traps of a different atomic structure which exist already before the high temperature $N_2$ anneal. They may be related to the presence of strain in the interfa- cial region (Gwyn 1969).

In addition to the deep hole traps also shallow hole traps occur in all Si -SiO₂ samples. They show a very high trapping efficiency at liq. $N_2$ tempe- rature. Little is known about their nature. There are indications that at least a substantial fraction of the holes captured in shallow traps are transfered to deep traps (Choi et al. 1983).

Fig. 6 (Singh et al. 1983a) schematically and tentatively shows the diffe- rent intrinsic traps and their proposed correlations. Also indicated is their distribution across the oxide film. The correlation between the cen- ters in fig. 6 and the structures shown in fig. 5 has not been cleared yet.

## 7. Radiation Damage

Upon exposure of oxidized silicon to electromagnetic radiation with energy larger than the SiO₂ bandgap (9eV) three effects are observed (see review by Ma, 1981): The oxide becomes positively charged due to capture of holes near the Si-SiO₂ interface upon generation of electron-hole pairs. Also, the density of the interface states increases; this effect will be dealt with in the next section. Both oxide charge and interface states can be easily annealed out at temperatures around 400°C. Finally, during irradia- tion neutral electron traps are being generated in the bulk of the oxide. They require annealing temperatures over 550°C for removal (Aitken et al.

1978, Aitken 1979). Ion bombardment will additionally lead to diplacement damage.

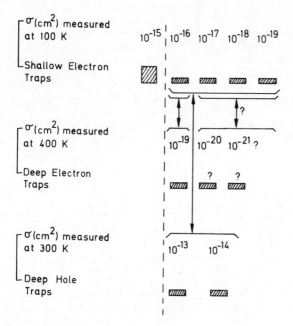

Fig. 6 Relation between electron and hole traps;
▨ in bulk, ▨ near Si-SiO$_2$ interface

The generation of neutral electron traps in principle can take place by two mechanisms (Gdula 1979): breaking of bonds during the excitation of electron-hole pairs and ionisation of bonds by the carriers created in the first process. The first process most likely would affect any bond in the oxide with equal probability, particularly for irradiation energies exceeding the SiO$_2$ band gap. As long as the two resulting parts do not move away from each other the original bond may reform again upon capture of an electron. However, particularly in the strain gradient near the Si-SiO$_2$ interface the broken bonds may diffuse away from each other (Grunthaner et al. 1982a). The second mechanism, involving electrons and holes of lesser energy, may via carrier capture at defects (strained bonds, water related defects) lead to local relaxation of the oxide structure and thereby to new electrically active centers. However, at the present time evidence supporting such detailed models still is lacking.

In a study of x-ray damage in MOS samples with Al electrode Schmitz and Young (1983) found after a 450°C anneal two deep neutral electron traps: one uniformly distributed across the oxide with $\sigma = (1-2) \times 10^{-16}$ cm$^2$ and density $(1-2) \times 10^{11}$ cm$^2$ after 1 Mrad (Si) irradiation; the second with $\sigma = (1-2) \times 10^{-15}$ cm$^2$ and somewhat lower density concentrated near the Si-SiO$_2$ interface. N$_2$ anneal for 2 hours at 1000°C before irradiation reduces the density of the subsequently formed $10^{-16}$ cm$^2$ center, whereas the $10^{-15}$ cm$^2$ center disappears. This behavior suggests that at least for the generation of the $10^{-15}$ cm$^2$ center the presence of some water related species may play a role since these are removed in a comparable time (fig. 4). A further reduction in the generation of neutral traps upon irradiation is obtained by a brief (30 sec) O$_2$ anneal following the N$_2$ anneal (Aslam and Balk 1983). Since this O$_2$ treatment removes the O-deficiency of the oxide apparently also stretched Si-Si bonds or trivalent Si centers (fig. 5) play a role in the generation of neutral traps.

In earlier studies on the radiation hardness of MOS systems in most cases the major concern was the negative shift in flatband voltage upon irradiation. For this purpose the contributions of the charge trapped in interface states and the trapped hole charges were lumped together; in such measurements the neutral electron traps play only a minor role because of

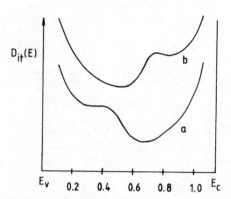

Fig. 7 Typical interface trap spectra after a) electron and b) hole injection

their lower density and smaller capture cross sections. Derbenwick and Gregory (1975) have reported that annealing in $N_2$, necessary to reduce the fixed positive oxide charge, should be carried out well below 1000°C, since otherwise a serious degradation of the radiation hardness is obtained. As shown in the previous section such annealing treatments led to an increase in the density of hole traps. The use of a 1100°C dry $N_2$ post oxidation anneal or $As^+$ implantation followed by a 875°C $N_2$ anneal, two treatments leading to O-deficient $SiO_2$, are mentioned by Grunthaner et al. (1982a) as methods for preparing "soft" gate oxides.

Electron traps may be generated by injection of hot electrons into the $SiO_2$ film. Data by Badihi et al. (1982) suggest that for a 70nm thick film already at a field of 3 MV $cm^{-1}$ a measurable rate of trap generation is obtained. It is unlikely that at these low fields electron hole pairs could be formed. The formation of trivalent Si species in the $SiO_2$ near the Si-$SiO_2$ interface upon injection of electrons or holes has indeed been observed by Grunthaner et al. (1982b) during in situ XPS measurements.

## 8. Generation of Interface States

It is generally agreed upon that interface states arise from the presence of strained or chemically unsaturated bonds at the Si-$SiO_2$ interface, (Sakurai and Sugano 1981, Flietner 1982). Typical examples of distributions of interface states over the forbidden gap of Si are shown in fig. 7. They consist of a U-shaped background which sometimes has some specific features superimposed upon it. For a review of the literature up to early 1981 the reader is refered to a recent review by Balk and Klein (1982). The so-called slow states were extensively studied first by Breed (1974, 1975). Their atomic structure is not understood at the present time. However, since they are often observed in samples which also show fast states, there may be some structural relation between the two types of centers.

The generation of interface states requires the flow of charge carriers through the oxide or the generation of electron-hole pairs in the oxide by electromagnetic radiation. Even though considerable progress has been made in recent years towards illuciding the mechanisms involved in the generation processes several significant questions remain unanswered. An important insight is that, at least for radiation induced surface states, but probably also for generation by Fowler-Nordheim or avalanche injection, the built-up is a two-stage process (Winokur and Sokolowski 1976, Winokur et al. 1979).

Upon irradiation electron-hole pairs are being generated throughout the oxide. According to McLean (1980) this step is independent of the polarity of the gate voltage; however, for very small fields the carriers will recombine again. The second step requires a positive gate voltage for the production of interface states. The rate of this step depends on the magnitude of the field and is thermally activated. For these reasons it was proposed that the second step involved the field assisted diffusion of a positive ion, possibly H$^+$, towards the interface. Here it would react immediately with weaker bonds, for example with Si-H, to form Si· and H$_2$ or with Si-OH to form Si· and H$_2$O (Revesz 1977, Svensson 1978). However, upon using strongly absorbing radiation, thereby creating electron-hole pairs away from the Si-SiO$_2$ interface, generation of interface states is observed when applying a negative voltage to the gate. This was explained by the assumption that neutral species like excitons (Weinberg and Rubloff 1978) or hydrogen atoms (Weinberg et al. 1979) were the moving species.

The presence of a positive gate voltage is not an absolute requirement for the generation of interface states using high energy radiation. It was demonstrated by Boesch (1982) that in samples with thick SiO$_2$ layer and poly-Si gate both in the presence of a very small negative and positive field interface states were obtained, be it at considerably lower concentrations. The generation rate at 77 K and 298 K is nearly equal, which shows that hole transport does not play a role (Bluzer et al. 1981). The presence of at least a small field helps to separate electrons and holes before they recombine. In this case only slow states were generated. Directly observed by Grunthaner et al.(1982a) was the formation and motion of trivalent Si defects in the SiO$_2$ upon irradiation with electrons. These defects move towards the silicon in the strain gradient near the Si-SiO$_2$ interface and give rise to interface states. Recent experiments by Chin and Ma (1983) appear to support this relation between interfacial stress prior to irradiation and the density of interface states being generated.

The study of the generation of interface states by Fowler-Nordheim tunnel injection of electrons shows that states are obtained for both polarities of gate voltage. However, at positive gate voltages fewer injected carriers are required to produce an interface state than at negative voltages (Hofmann and Dorda 1981, DoThanh and Balk 1983). This behavior appears to indicate that energetic electrons are able to create interface states; however, the holes obtained as a "by product" from impact ionization in Fowler-Nordheim injection are more efficient in creating defects.

Since radiation experiments and Fowler-Nordheim injection both produce electrons and holes in the oxide one expects similar effects regarding the creation of interfacial defects for the two methods. This is qualitatively the case. However, for tunnel injection the ratio of electrons to holes is considerably larger than one whereas energetic radiation generates equal numbers of both carriers. It was shown by Knoll et al.(1982a, 1982b, 1983) that by correcting for the neutralisation of trapped holes by electrons and for subsequent detrapping of these electrons by impact ionization agreement between irradiation and tunnel injection results is obtained. Thus the captured holes (in the case of tunnel injection the density of trapped holes, whether they have been neutralized by captured electrons or not) dominate the generation of interfacial defects. In fact, Hu and Johnson (1980) showed in a Fowler-Nordheim injection experiment at positive bias where they separated hole capture (at 77 K) and generation of interface states (which is slow at liq. N$_2$ temperature but takes place faster at room temperature or above) that each trapped hole in a slow process (one year at room temperature) gives rise to a surface state.

Photo injection and avalanche injection yield data that are easier to analyse, since only one type of carrier is involved. Avalanche injection of electrons at room temperature produces first capture of electrons in bulk states and is followed by a turn around in the direction of change of the flatband voltage due to the built-up of positive charge in slow states near the Si-SiO$_2$ interface (Gdula 1976, Young et al.1979). At the same time fast interface states are being generated. The rate of generation appears to correlate with the density of water related electron traps in the oxide (Miura et al.1980, Feigl et al.1981, Lai and Young 1981, Fischetti et al. 1982 a,b). Using photoinjection of electrons, again at low field (1-2 MV cm$^{-1}$), Pang et al.(1980, 1982) observed that the generation of interfacial states between 90 and 273 K is independent of temperature and of the direction of the current. However, the generation rate increases nearly linearly with the oxide thickness and sublinearly with the oxide field. The "yield" of interface states per electron is extremely low ($10^{-4}$ to $10^{-7}$). Also in this case there is a direct relationship with the density of water related traps (Zekeriya and Ma 1983). Since the migration of atoms or atomic groups would be temperature dependent the authors conclude that most likely only the stress at the interface is involved in the formation of the centers.

Avalanche injection of holes also leads to the development of interface states (Lai 1981, 1983). In this case the yield is quite high (1 to $10^{-3}$ states per hole). DoThanh et al.(1983) established that the yield decreases distinctly with decreasing temperature. At 298 K, for each trapped hole an interface state was obtained, at 77 K considerably less; this behavior suggests that only holes captured in deep traps produce interface states. The generation rate clearly correlated with the density of deep hole traps; it increases with high temperature annealing, decreases with an O$_2$ treatment.

An extensive study of the energy spectrum of the interface states obtained by exposure to 10,2 eV photons was made by Sah et al.(1982, 1983). In addition to a constant background and a U-shaped distribution the spectrum shows two peaked features (most likely caused by the presence of some hydroxyl species) at 0.75-0.85 eV above the valence band edge and one at 0.95 eV above the valence band edge. The double peak at 0.75-0.85 eV is commonly known as the radiation induced peak (Ma et al. 1975). It has also been reported to appear upon avalanche injection of holes, its density increasing with that of the intrinsic hole traps; a further increase in this peak and a new feature at 0.4 eV above the valence band appear upon subsequent tunnel injection of electrons (DoThanh et al. 1983). On the other hand, Lai (1981, 1983) obtains only the first peak and exclusively upon photo injection of electrons after avalanche injection of holes. Upon avalanche injection of electrons DoThanh et al.(1983) obtain, in addition to the U-shaped profile, the peak at 0.4 eV.

It thus appears that the literature data on the generation of interface states still contain several discrepancies. Part of these may be due to the fact that the experiments were carried out with poorly characterized samples. However, the conclusion that the rate of generation is directly related to the density of trapping centers for the injected carrier in the film is beyond doubt.

## 9. Dielectric Breakdown

Because of the high fields occuring in short-channel devices the dielectric strength of the gate insulator has become a matter of particular concern. Continued Fowler-Nordheim injection of electrons in SiO$_2$ will lead to reduced sample resistance; a current instability arises which may ultimately lead to breakdown (Osburn and Weitzman 1972). Quite generally, charge trapping is being observed before local destruction of the film takes place. However, notwithstanding the considerable practical interest of this problem the breakdown of thin insulators is still "an open subject" (Klein 1983). For films at least 10nm thick electron-hole pairs may be generated followed by trapping of the holes; upon continued injection neutralisation of the trapped holes by electrons and subsequent impact ionization will additionally take place, as discussed in the preceding section. The positive charge lowers the injection barrier, which leads to increased current at a given stressing voltage. However, also avalanche injection of holes leads in samples with high initial intrinsic hole density already for relatively small amounts of injected holes to considerable capture of positive charge and dielectric breakdown (Aslam and Balk 1983). It is not clear if in this case impact ionisation plays a role.

It thus appears to be important to minimize the density of hole traps in the $SiO_2$ layer. In agreement with this concept Aslam et al. (1983) found that a prolonged high temperature $N_2$ anneal led to an increased incidence of low-field breakdown for both gate polarities in samples with Al and poly-Si electrodes. A brief anneal in $O_2$ completely suppresses these events (fig. 8). The similarly improved behavior obtained upon a reoxidation treatment of $N_2$ annealed films by Cohen (1983) appears to support this idea. The same explanation most likely applies for the extreme stability and good breakdown strength of oxidized-nitridized oxide films for EEPROM applications (Chang et al. 1982).

For very thin oxides (10nm)and below there appears to be a pronounced increase of the breakdown field (Harari 1978, Adams et al. 1980). Electron injection in this case generates electron traps. By cooling down to 77 K during injection an appreciable reduction of trap generation and increase in dielectric strength is attained (Harari 1978). It is interesting to note that also the generation of interface states by avalanche injection of electrons and holes is slowed down considerably upon cooling (DoThanh et al. 1983). This behavior appears to indicate that in both cases a thermally activated process is involved. Maserjian and Zamani (1982) report the formation of positively charged traps upon Fowler-Nordheim injection of electrons in very thin (approx.5nm) $SiO_2$ films; in this case the energy of the carriers was definitely below that of the bandgap. These states are comparable with the centers

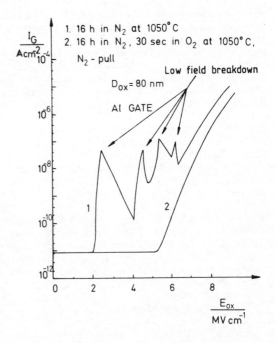

Fig. 8 Elimination of low field breakdown in $N_2$ annealed $SiO_2$ by $O_2$ treatment

obtained from the breaking of strained Si-O-Si bonds observed by XPS (Grun-thaner and Maserjian 1978). A remarkable breakdown strength and absence of low field breakdown is exhibited by very thin ($<$10nm) gate insulators pre pared by direct nitridation of the silicon substrate (Ito et al. 1982). How-ever, very little is known about the trapping behavior of such thin films and their interaction with injected carriers. A systematic study of these topics would be of considerable importance for the developemt of VLSI technology.

## 10. Implications for the Gate Oxide Technology

The discussion in the foregoing sections has identified three types of technological process steps which have a distinct effect on the trapping behavior of the oxide. These are: implantation through and diffusion into the oxide of the dopants As and P; the use of ionizing radiation and the application of high temperature annealing treatments to the system. A suc-cessful device fabrication process will have to take these different as-pects into account.

Because of its large capture cross section for electrons $As^+$ implantation and diffusion appears to have particularly harmful side effects. An extre-me example is the finding by Osburn et al. (1982) that immediately upon $As^+$ implantation of the source and drain in a self aligned poly-Si gate tech-nology shorting of the thin oxide at the thin oxide perimeter was observed. In this case the problem was overcome by using an oxide spacer, which is also useful in implementing the gradual junction approach and in incorpo-rating self-aligned silicides. Devices prepared using a channel implant through the gate oxide were found to be more susceptable to electron cap-ture in the case of $As^+$ than for $B^+$ implants (Bauer and Balk 1983). It thus appears advisable to avoid implantation through the gate oxide where possible.

Questions relating to radiation damage have obtained considerable impor-tance in the submicron technology. Takeda et al. (1982) showed an increase in trapping efficiency for electrons by over an order of magnitude when comparing metal (Al?) gate devices prepared by electron-beam lithography with devices of the same dimensions by photolithography. It appears that in this case the final annealing step (30 min $H_2$ at 450°C) is clearly not sufficient to remove the neutral centers. Similar results were reported by Chen et al.(1982). The required temperatures of 550°C can only be appli-ed in a poly-Si gate technology along with a source and drain contact me-tallurgy that is sufficiently stable at this temperature (for example, by the use of refractory metals). Utilization of e-beam resists with improv-ed sensitivity would permit lowering of the radiation dose and thereby re-duce the damage. A similar improvement could be expected from the masking action of thick multilevel resists. A last possibility of improvement is the use of radiation hardened oxide films. The preparation of such films requires annealing treatments at 900°C or higher (Terry et al. 1983).

The application of an annealing step in a neutral ambient is important for reducing the fixed positive oxide charge in the Si-$SiO_2$ system. Ideally, a temperature close to 1000°C should be used for this treatment. However, it is usually important to avoid O-deficiency in the films. To this end the $N_2$ annealing step followed by a brief $O_2$ treatment should be carried out before deposition of the poly-Si film in the case of a poly-Si or polycide gate technology.

The poly-Si and similarly the polycide (poly-Si metal disilicide) gate technology have the advantage that during gate deposition the water related traps are effectively removed. Also, they permit annealing of radiation in-duced neutral traps. However, the introduction of a silicide layer on top of the poly-Si generally will induce additional strain in the MOS system. Such strain makes the system more liable to the generation of interface (and possibly also bulk) states upon interaction with injected carriers. The passivation of the disilicides by thermal oxidation, by itself an at-tractive possibility, requires an additional high temperature step, for ex-ample in $H_2O/O_2$ at 900°C in the case of $TiSi_2$ (Valyi et al. 1983). Both oxi-dation ambient and temperature are not necessarily suitable to improve the trapping behavior of the MOS system.

Summarizing, it may be stated that a substantial amount of information has become available on the individual process steps and their effect in device properties. However, there are many open questions regarding the interac-tion between the steps that should receive high priority in answering them. Only this further information will permit formulation of an optimi-zed stable MOSFET technology.

As a final question we will briefly discuss the stability problems arising in the operation of MOSFET circuits and devices at reduced temperatures, e.g. 77 K. In addition to the improvement in speed during low temperature operation also all chemically activated processes are being suppressed. This concerns both processes like the wear-out of the oxide due to $Na^+$ drift and the drift of intrinsic defects in strain gradients. The effective-ness of interface state generation is in most cases considerably lower. On the other hand, the multiplication of charge carriers increases for lower temperatures and so does the trapping of holes and electrons in shallow states. Upon warm-up to room temperature only a fraction of these charges becomes detrapped. It thus may be expected that also the low temperature operation of MOSFETs will present problems that are at least as severe as those encountered at room temperature. In fact, Bauer and Balk (1982) found in a study on short-channel n-MOS ring oscillators that the supply voltage required for onset of oscillation at room temperature exhibited a larger shift when stressed at 77 K than at room temperature. This agrees with an earlier study by Ning et al. (1979) showing a larger decrease in the transconductance of n-channel MOSFETs at 77 K than at room temperature.

## 11. Conclusions

In the foregoing discussion we have shown that the generation of hot carri-ers (both electrons and holes) becomes increasingly pronounced for smaller channel length devices. Modifications of the doping profile and changes of the geometry of the device (offset gate) can only bring limited relief, particularly if one is not prepared to make sacrifices in the area of per-formance.

The technological methods and approaches used in realising the MOS gate sy-stem in the submicron technology lead in a number of cases to increased in-teraction between oxide and injected carriers; the oxide technology thus further contributes to the instability of the device characteristics.

Important are the avoidance of the dopant impurities (particularly As and also P) and of $H_2O$ related traps in the oxide, since their presence cau-ses hole and electron traps in the insulator. Particularly undesirable is the presence of dangling bond centers, like those related to O-deficiency of the insulator. These centers act as hole traps, but are also effective in the irreversible trapping of electrons at reduced temperatures. The use of the poly-Si and polycide technologies with refractory metal contact systems will permit removal of radiation damage. However, it is not clear that the polycide gate electrode does not affect the gate stability due to mechanical strain in the MOS system and thereby enhances carrier capture.

One of the important tasks for the immediate future is a systematic study of the interactions between the different technological parameters in their effect on device stability. The interesting nitridized oxide or nitridized silicon technology should be included in this analysis.

The miniaturization of MOSFET devices is likely to take place in a number of discrete steps. Experience shows that for each reduction of the critical parameters (particularly gate length) by 30-50% a new series of physical and technological problems arises. These have to be solved first before a next step towards a reproducible miniaturisation can be taken. It appears likely that a gate length reduction from 1 to 0.5 μm will occupy the industry for several years. In doing so hot carrier injection problems will still play an important role. Only in a next phase of development along with a reduction of the device operating voltages well below the Si-SiO$_2$ injection barrier of electrons (3.1 ev) will the hot carrier injection instabilities probably disappear (Tam et al. 1983).

## Acknowledgements

The author is indebted to R.F. DeKeersmaecker and G. Dorda for providing information on unpublished results and discussions. He likes to thank his coworkers M. Aslam, F. Bauer and L. DoThanh, whose support and criticism were very helpful in writing this review.

## References

Adams A C, Smith T E and Chang C C 1980 J. Electrochem. Soc. 127 1787
Aitken J M 1979 IEEE Trans. Electron Devices ED-26 372
Aitken J M and Young D R 1977 IEEE Trans. Nucl. Sci. NS-24 Dec 2128
Aitken J M, Young D R and Pan K 1978 J. Appl. Phys. 49 3386
Alexandrova S and Young D R 1983 J. Appl. Phys. 54 174
Aslam M and Balk P 1983 Insulating Films on Semiconductors eds J F Verwey
    and D R Wolters (Amsterdam: North Holland Publ. Co)
Aslam M, Balk P and Young D R 1983 to be published
Badihi A, Eitan B, Cohen I and Shappir J 1982 Appl. Phys. Lett. 40 396
Balk P 1974 Solid State Devices, 1973 (London: The Inst. of Physics Conf.
    Series No 19) p 51
Balk P 1975 J. Electron. Mat. 4 635
Balk P 1983 Insulating Films on Semiconductors ed J F Verwey and D R Wol-
    ters (Amsterdam: North Holland Publ. Co)
Balk P and Klein N 1982 Thin Solid Films 89 329
Bauer F and Balk P 1982 presented at ESSDERC 82
Bauer F and Balk P 1983 presented at ESSDERC 83
Bennett A J and Roth L M 1971 J. Phys. Chem. Solids 32 1251
Bhattacharyya P, Manchanda L and Vasi J 1982 J. Electrochem. Soc. 129 2772
Bluzer N, Affinito D and Blaha F 1981 IEEE Trans. Nucl. Sci. NS-28 Dec 4074
Boesch Jr H E 1982 IEEE Trans Nucl. Sci. NS-29 Dec 1446
Borchert B, Hofmann K R and Dorda G 1983 to appear in Electron. Lett.
Breed D J 1974 Solid State Electron. 17 1229
Breed D J 1975 Appl. Phys. Lett. 26 116

Chang T T L, Jones H S, Lee J, Ho C, Lai S K and Dham V 1982 IEDM 82 Digest of Technical Papers p. 810

Chen J Y, Henderson R C, Patterson D O and Martin R 1982 IEEE Electr. Dev. Lett. EDL-3 13

Chin M R and Ma T P 1983 Appl. Phys. Lett. 42 883

Choi S C, Aslam M and Balk P 1983 to be published

Cohen S S 1983 J. Electrochem. Sco. 130 929

Cottrell P E, Troutman R R and Ning T H 1979 IEEE Trans. Electron Devices ED-26 520

DeKeersmaecker R F, DiMaria D J and Pantelides S T 1978 The Physics of SiD$_2$ and its Interfaces ed S T Pantelides (New York: Pergamon Press) p. 189

DeKeersmaecker R F and DiMaria D J 1980a J. Appl. Phys. 51 532

DeKeersmaecker R F and DiMaria D J 1980b J. Appl. Phys. 51 1085

DeKeersmaecker R F 1983 Insulating Films on Semiconductors ed J F Verwey and D R Wolters (Amsterdam: North Holland Publ. Co)

Derbenwick G F and Gregory B L 1975 IEEE Trans. Nucl. Sci. NS-22 Dec 2151

DiMaria D J 1978 The Physics of SiO$_2$ and its Interfaces ed S T Pantelides (New York: Pergamon Press) p. 160

DoThanh L, Aslam M and Balk P 1983 presented at ESSDERC 83

DoThanh L and Balk P 1983 Insulating Films on Semiconductors ed J F Verwey and D R Wolters (Amsterdam: North Holland Publ. Co)

Fair R B and Sun R S 1981 IEEE Trans. Electron Devices ED-28 83

Feigl F J, Young D R, DiMaria D J, Lai S K and Calise J 1981 J. Appl. Phys. 52 5665

Fischetti M V, Gastaldi R, Maggioni F and Modelli A 1982 a,b J. Appl. Phys. 53 3129, 3136

Flietner H 1982 Defect Complexes in Semiconductor Structures eds J Giber, F Beleznay, J C Czép and J Lázló (Berlin: Springer Verlag) p. 247

Gdula R A 1976 J. Electrochem. Soc. 123 42

Gdula R A 1979 IEEE Trans. Electron Devices ED-26 644

Gesch H, Leburton J-P and Dorda G E 1982 IEEE Trans. Electron Devices ED-29 913

Griscom D L and Fowler W B 1978 The Physics of SiO$_2$ and its Interfaces ed S T Pantelides (New York: Pergamon Press) p. 97

Grunthaner F J and Maserjian J 1978 The Physics of SiO$_2$ and its Interfaces ed S T Pantelides (New York: Pergamon Press) p. 389

Grunthaner F J, Grunthaner P J and Maserjian J 1982a IEEE Trans. Nucl.Sci. NS-29 Dec 1483

Grunthaner F J, Lewis B F, Maserjian J and Madhukar A 1982b J. Vac. Sci. Tech. 20 747

Gwyn C W 1969 J. Appl. Phys. 40 4886

Hara H, Okamoto Y and Ohnuma H 1970 Jap. J. Appl. Phys. 9 1103

Harari E 1978 J. Appl. Phys. 49 2478

Hartstein A and Young D R 1981 Appl. Phys. Lett. 38 631

Hofmann K R 1983 to appear in Proc. of INFOS 83

Hofmann K R and Dorda G 1981 Insulating Films on Semiconductors eds M. Schulz and G Pensl (Berlin: Springer Verlag) p. 122

Hofstein S 1967 Solid State Electron. 10 657

Hu G J and Johnson W C 1980 Appl. Phys. Lett. 36 590

Irene E A, Tierney E and Angilello J 1982 J. Electrochem. Soc. 129 2594

Ito T, Nakamura T and Ishikawa H 1982 IEEE Trans. Electron Devices ED-29 498

Kamata T, Tanabashi K and Kobayashi K 1976 Jap. J. Appl. Phys. 15 1127

Klein N 1983 Thin Solid Films 100 335

Knoll M, Bräunig D and Fahrner W R 1982a J. Appl. Phys. 53 6946

Knoll M, Bräunig D and Fahrner W R 1982 b IEEE Trans. Nucl. Sci. NS-29

Dec 1471

Knoll M, Bräunig D and Fahrner W R 1983 Insulating Films on Semiconductors eds J Verwey and D Wolters (Amsterdam: North Holland Publ. Co)

Lai S K 1981 Appl. Phys. Lett. 39 58

Lai S K 1983 J. Appl. Phys. 54 2540

Lai S K and Young D R 1981 J. Appl. Phys. 52 6231

Lai S K, Young D R, Calise J A and Feigl F J 1981 J. Appl. Phys. 52 5691

Lander J J and Morrison J 1962 J. Appl. Phys. 33 2089

Liang M-S and Hu C 1981 IEDM 81 Digest of technical papers p. 396

Ma T P 1981 Semiconductor Silicon 1981 eds H R Huff, R J Kriegler and Y Takeishi (Princeton N J: The Electrochem. Soc.) p. 427

Ma T P, Scoggan G and Leone R 1975 Appl. Phys. Lett. 27 61

Marczewski M and Strzalkowski I 1982 Appl. Phys. A 29 233

Maserjian J and Zamani N 1982 J. Vac. Sci. Technol. 20 743

McLean F B 1980 IEEE Trans. Nucl. Sci. NS-27 Dec 1651

Miura Y, Yamabe K, Komiya Y and Tarni Y 1980 J. Electrochem. Soc. 127 191

Mizutani Y, Taguchi S, Nakahara M and Tango H 1981 IEDM 81 550

Ng K K and Taylor G W 1982 Device Research Conference Ft Collins, Co

Ngai K L and White C T 1981 J. Appl. Phys. 52 320

Nicollian E H, Goetzberger A and Berglund C N 1969 Appl. Phys. Lett. 15 270

Nicollian E H, Berglund C N, Schmidt P F and Andrews J M 1971 J. Appl. Phys. 42 5654

Ning T H 1978 J. Appl. Phys. 49 5997

Ning T H, Osburn C M and Yu H-N 1977 J. Electron. Mater. 6 65

Ning T H, Cook P W, Dennard R H, Osburn C M, Schuster S E and Yu H-N 1979 IEEE Trans. Electron Devices ED-26 346

Ogura S, Tsang P J, Walker W W, Critchlow D L and Shepard J F 1980 IEEE Trans. Electron Devices ED-27 1359

Osburn C M, Cramer A, Sweighart A M and Wordeman M R 1982 Extended Abstracts ECS Meeting Oct 1982 (Pennington N J: The Electrochem. Soc.) 82-2 278

Osburn C M and Weitzman E J 1972 J. Electrochem. Soc. 119 604

Pang S, Lyon S A and Johnson W C 1980 The Physics of MOS Insulators eds G. Lucovsky, S T Pantelides and F L Galeener (New York: Pergamon Press) p. 285

Pang S, Lyon S A and Johnson W C 1982 Appl. Phys. Lett. 40 709

Pantelides S T 1982 Thin Solid Films 89 103

Revesz A G 1973 J. Non-Cryst. Solids 11 309

Revesz A G 1977 IEEE Trans. Nucl. Sci. NS-24 Dec 2102

Revesz A G 1979 J. Electrochem. Soc. 126 122

Sah C-T, Sun J Y-C and Tzou J J-T 1982 J. Appl. Phys. 53 8886

Sah C-T, Sun J Y-C and Tzou J J-T 1983 J. Appl. Phys. 54 2547

Sakurai T and Sugano T 1981 J. Appl. Phys. 52 2889

Schmitz W and Young D R 1983 to be published

Singh R, Aslam M and Balk P 1983a to be published

Singh R, Maier M, Kräutle H, Young D R and Balk P 1983b to be published

Smeltzer R K 1982 IEEE Trans Nucl. Sci. NS-29 Dec 1467

Svensson C M 1978 $SiO_2$ and its Interfaces ed S T Pantelides (New York: Pergamon Press) 328

Takeda E, Kume H, Toyabe T and Asai S 1982 IEEE Trans. Electron Devices ED-29 611

Takeda E, Nakagome Y, Kume H and Asai S 1983 IEE Proc Pt I 130 144

Tam S, Hsu F-C, Hu C, Muller R S and Ko P K 1983 IEEE El. Dev. Lett. EDL-14 249

Terry F L, Aucoin R J, Naiman M L and Senturia S D 1983 IEEE Electron Dev. Lett. ELD-4 191

Vályi G, Lu Z, Maier M, Steffen A and Balk P 1983 presented at ESSDERC 83
Weinberg Z A and Rubloff G W 1978 Appl. Phys. Lett. $\underline{32}$ 184
Weinberg Z A, Young D R, DiMaria D J and Rubloff G W 1979 J. Appl. Phys. $\underline{50}$, 5757
Winokur P S and Sokolowski M 1976 Appl. Phys. Lett. $\underline{28}$ 627
Winokur P S, Boesch Jr H E, Mc Garrity J M and McLean F B 1979 J. Appl. Phys. $\underline{50}$ 3492
Woods M H and Williams R 1976 J. Appl. Phys. $\underline{47}$ 1082
Young D R 1981 J. Appl. Phys. $\underline{52}$ 4090
Young D R, Irene A E, DiMaria D J, DeKeersmaecker R F and Massoud H Z 1979 J. Appl. Phys. $\underline{50}$ 6366
Zekeriya V and Ma T P 1983 Appl. Phys. Lett. $\underline{43}$ 95

*Inst. Phys. Conf. Ser. No. 69*
*Paper presented at ESSDERC/SSSDT 1983, Canterbury 13–16 Sept. 1983*

# Radiation effects in semiconductors: technologies for hardened integrated circuits

Jean-Marie CHARLOT
Service Electronique, Commissariat à l'Energie Atomique
BP 12 - 91680 Bruyères-le-Chatel , FRANCE

Abstract.

Various technologies are used to manufacture integrated circuits for electronic systems. But for specific applications, including those with radiation environment, it is necessary to choose an appropriate technology or to improve a specific one in order to reach a definite hardening level.
The aim of this paper is to present the main effects induced by radiation(neutrons and gamma rays) into the basic semiconductor devices, to explain some physical degradation mechanisms and to propose solutions for hardened integrated circuit fabrication.
The analysis involves essentially the monolithic structure of the integrated circuits and the isolation technology of active elements.
In conclusion, the advantages of EPIC and SOS technologies are described and the potentialities of new technologies (GaAs and SOI) are presented.

## 1. Introduction.

Integrated circuits are widely used in the design of logic and memory functions of electronic systems. They are particularly well adapted to solve miniaturization and low power consumption problems. For specific military, space and even civil (nuclear industrial field) applications, their use is conditionned by their ability to work under high level radiation environment.
Degradations are induced by this environment into materials and components and are dependent on incident radiation type, semiconductor characteristics and time.
The nuclear environment taken into account here involves essentially:
- a neutron burst with fission and fusion neutrons
- an ionizing X and gamma radiation pulse, the energy of which is about 1 MeV.

Available technologies to manufacture integrated circuits are divided in two families: The bipolar family, where the junction transistor is the basic element and the unipolar family with the MOS transistor (Metal-Oxide-Semiconductor).
After having described the main radiation effects induced in these two types of components, their influence upon the integrated circuits of various technologies will be analysed and actions to improve their behaviour under radiation will be discussed.

2. Interactions and damages in semiconductors.

Radiation effects are generally divided in two classes taking into account
the energy deposition process inside the matter; energy can be transferred
either to electrons (ionization) or to nucleus (atomic displacements).
Ionization is the process of removing electrons from their parent atoms
and thereby positive ions and free or unbound electrons are created.
Free electrons can be due to charged particles (electrons,protons, recoil
atoms...) and to photons, directly or not according to three types of interactions: photoelectric and compton effects and pair production. The figure 1 shows the relative importance of the three types of photon interactions in function of absorbing material and photon energy.

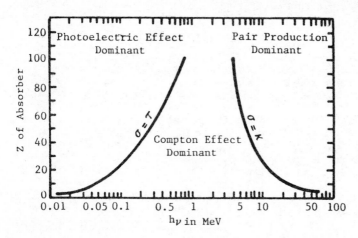

Fig. 1

τ = photoelectric
    cross-section

σ = Compton cross-
    section

κ = pair-production
    cross-section

Displacement effects are due to the interaction  of charged particles or
neutrons with the nucleus. Recoil atom so created possesses enough energy
to collide with other lattice atoms and to induce others displacements in
cascade during its slowing down.

Electronic disturbances lead to macroscopic effects the more important of
which are:
- the build up of a positive charge in the oxides and specially in the
  SiO layers
- the creation, during ionization, of positive and negative charge carriers into semiconductor which can move by diffusion or conduction to
  induce photocurrents.

Atomic perturbations (density and spatial distribution of displaced atoms
depend upon incident particle type and energy) give rise to common defects
(vacancies and interstitials), unstable and chemically reactive. The final
stable state of semiconductor is characterized by the existence of donor
and acceptor levels into the forbidden gap. This modified state has conse-
quences on fundamental parameters of active elements.

In qualitative terms, the degradations of the semiconductor components are
proportionnal to the radiation absorbed dose in Gy(Si)* or to the radiation
fluence in particles/cm  and lead to permanent dammage. The defects and the
associated perturbations are defined as transient or persistent when the radiation is time-dependent   and   the induced effects end   with the cause
or more slowly than the cause, respectively.

* 1 Gy = 100 Rad

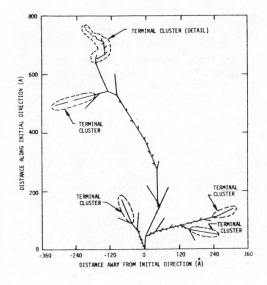

Fig. 2 Defect generation
Picture of typical recoil-
atom track, with primary
energy of 50 KeV.

after V.A.Van Lint and
                    R.E.Leadon.

Fig. 3 Damage function
       (Silicon)

after E.C.Smith
and J.Sauret.

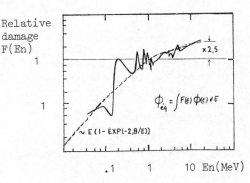

Relative
damage
F(En)

$\varphi_{eq} = \int F(E)\,\varphi(E)\,dE$

$\sim E\,(1 - EXP(-2,8/E))$

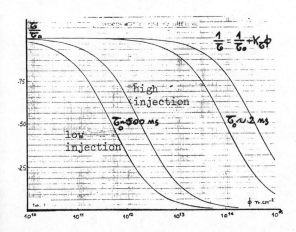

$$\frac{1}{\tau} = \frac{1}{\tau_0} + K_\tau \phi$$

high
injection

$\tau_0 \simeq 500\ ms$        $\tau_0 \simeq 1,2\ ms$

low
injection

Fig. 4

Typical relative varia-
tions of lifetimes

3. Neutron effects.

Neutrons are characterized by their energy spectrum ( > 10 KeV) and their fluence measured in $n/cm^2$.

The relation is not direct between the initial phenomenon, the neutron-semiconductor material (Si) interaction, and the final consequence, the modification of the component electrical performance; several steps have to be considered.

3.1. Displacement damage. [1-4]

The displacement of a silicon atom from its lattice position requires about 25 eV. An incident neutron with a sufficent energy arriving inside the lattice can displace one atom, delivering to it a part of its energy in an elastic collision. This displaced atom (recoil atom) and incident neutron can displace again other atoms if their energies are still sufficent.

The effects of displaced atoms in a silicon lattice are due to the creation of simple defects and defect clusters (figure 2). The number of defects created by the recoil atom and the dimensions of the clusters depends on the neutron energy (about 1000 displacements for one 1 MeV neutron).

It is interesting to note that the dimensions of a cluster is comparable to the dimensions of an integrated circuit basic transistor (   1 micron for one 14 MeV neutron).

The damage function into the silicon versus the neutron energy  is displayed in figure 3. One can see that the displacement effects become significant for En = 100 KeV and for En = 14 MeV they are about three times more important than for En = 1 MeV. Usually, the neutron fluence is expressed in equivalent 1 MeV fluence via the damage function ( silicon ).

3.2. Electronic induced effects.

The defects created by atom displacements act as carrier recombination centers and therefore modify the silicon properties.

Minority carrier lifetime.
The relation between neutron fluence and minority carrier lifetime is:

$$\Delta \left( \frac{1}{\tau} \right) = \frac{1}{\tau_\Phi} - \frac{1}{\tau_0} = K_\tau \Phi$$

$\tau_0$ is the preirradiated lifetime
$\tau_\Phi$ is the postirradiated lifetime
$K_\tau$ is the lifetime damage constant
(in $cm^2 s^{-1} n^{-1}$)
$\Phi$ is the neutron fluence

K is dependent on the type and impurity content of the semiconductor, injection level, neutron spectrum ... (figure 4).
An average value of K for silicon is about $K \simeq 10^{-6} cm^2 s^{-1} n^{-1}$.

Majority carrier density.
Defects result in the introduction of allowed energy states within the forbidden gap of silicon, reducing the number of carriers in the conduction band.

The variations of majority carrier concentrations in silicon can be written empirically according to Buelher [5]

$$N = No\ e^{-\frac{\Phi}{444\ No}0.77}$$

$$P = Po\ e^{-\frac{\Phi}{387\ Po}0.77}$$

Majority carrier mobility.
The semiconductor mobility is determined by collisions of carriers with vibrating lattice atoms and long range coulomb collisions with ionized donor and acceptor atoms. In increasing the number of donor and acceptor defects, the number of collisions is therefore increased and the mobility decreases following the relation

$$\Delta\left(\frac{1}{\mu}\right) = \frac{1}{\mu_\Phi} - \frac{1}{\mu_0} = K_\mu \Phi$$

$$K_\mu \simeq 10^{-19} V^{-1} s^{-1} n^{-1}.$$

3.3. Neutron effects on bipolar transistors.

All of the above effects may be important, but the one that is often of most concern in bipolar transistors is the decrease of the current gain produced by the reduction in minority lifetime. An order of magnitude decrease in current gain, say from 100 to 10, may be observed in transistors that have been exposed to a neutron fluence of $10^{14}$ n/cm$^2$.

The reciprocal of transistor gain ($h_{FE}$) is a function of neutron fluence as:

$$\Delta\left(\frac{1}{h_{FE}}\right) = \frac{1}{h_{FE\Phi}} - \frac{1}{h_{FEO}} = t_b K' \Phi$$

where $h_{FEO}$ is the preirradiated gain
$h_{FE\Phi}$ is the postirradiated gain
$t_b$ is the base transit time
$K$ is the gain damage constant
$\Phi$ is the neutron fluence

Note that the base transit time is inversely proportionnal to the gain bandwith product $F_T$ and hence $\Delta(1/h_{FE}) = K'' \Phi / F_T$ .

It is also possible to place the emitter current into this relation in order to show that the gain degradation is a function of the injection level.

$$\Delta(1/h_{FE}) = K''' \Phi / F_T . I_E^{0.3}$$

The increase of the collector resistivity ($\rho \propto 1/N\mu$) causes the breakdown voltage and VCE(SAT) to increase.
The breakdown voltage increase is here an improvement over the pre-irradiation value. The corresponding increase in VCE(SAT), however, is undesirable since VCE(SAT) is an important parameter in logic and amplifier circuitry.

In conclusion, transistors with good resistance to neutron fluence will have very thin bases and shallow emitters, with high $F_T$. It will be necessary to minimize the collector resistivity and to adjust the base doping.

### 3.4. Neutron effects on MOS transistors.

MOS transistors, wether N-orP-channel are majority carrier devices and for that, not susceptible to neutron irradiation below $10^{13}$ n/cm$^2$. Ionizing effects associated with neutrons at high level fluence will be taken into account with the ionizing total dose.

### 4. Photon effects.

Photons, as neutrons, are indirectly ionizing; but on account of the pulsed character of photon radiation considered, the transient and persistent effects are more varied.

For high energy photons ($\simeq$ 1 MeV) Compton effect is directional; it can generate a macroscopic electron current (Compton current) with which is associated an electromagnetic field. This      is able to induce particular disturbances in the component behaviour.

For lower energy photons (1 KeV - 200 KeV) photoelectric effect is predominant. It gives rise to important charge variations at the interfaces.

The generation of a positron-electron pair becomes predominant when the photon energy is of several MeV.

Under gamma or X rays (and neutrons above 1 MeV) , free carriers generated can -undergo random motion
-diffuse to regions of lower carrier concentration
-drift with applied electric field
-be trapped at impurity atom sites
-recombine with their mates.

### 4.1. Total dose effects.

The main effects due to ionizing total dose are the buildup of positive charg in the oxide  layer, an increase of electronic states at the interface Si-SiO$_2$ and the creation of electron traps into the oxide.

### 4.1.1. Bipolar transistors.

For diodes and transistors, positive charges induced into the protection oxide can invert the doping of the underlying silicon layer, giving rise to leakage currents between the two parts of the junction or into the base of a transistor. This last effect reduces the transistor gain. (figure 5)

But now, with good oxides, these effects are minimized so that no influence is seen below  doses equivalent to $10^4$ Gy(Si). It is not the same thing with MOS technologies.[7]

Fig. 5

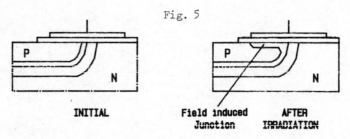

INITIAL                    Field induced        AFTER
                            Junction          IRRADIATION

4.1.2. <u>MOS transistors</u>.

Charge buildup into the gate oxide is the principal reason of the sensitivity of MOS transistors to an ionizing radiation. Under radiation, free e-
lectrons and holes are created into the oxide which move with the applied
voltages. Electrons are mobile and can usually be swept out of the $SiO_2$
while the holes, less mobile, are trapped in the gate insulator. The result
is a net positive space charge which induces negative charges in the underlying silicon which modify the electronic characteristics of the semiconductor in the channel region. [6]

The basic electrical parameter that measures this effect is the threshold
voltage ( $V_T$ ).

In addition to the trapped charges, electronic states may be created at the
silicon-oxide interface. In p-type silicon (N-channel transistor) , the interface states are charged neutral or negative; they can compensate the positive chargein the insulator. In N-type silicon (P-channel transistor) the
interface states are neutral or positive; they tend to add to the positive
oxide charge.

The contribution of these radiation effects can be introduced into the expression of the transistor threshold voltage which is before radiation

$$V_T = - \frac{Q_{ox} + Q_{ss}}{C_{ox}} + \Phi_{MS}$$

where $Q_{ox}$ is the oxide charge
$Q_{ss}$ is the **interface** state charge
$C_{ox}$ is the gate capacitance
$\Phi_{MS}$ is the silicon-metal work function

and which becomes after radiation

$$V_T = - \frac{(Q_{ox} + Q_{rsc}) + (Q_{ss} + Q_{rss})}{C_{ox}} + \Phi_{MS}$$

where $Q_{rsc}$ is the radiation induced positive
charge

$Q_{rss}$ is the radiation induced interface
state charge

The figure 6 shows the shifts of the transfer curves ($I_D$-$V_G$) for N-and P-
transistors. It can be seen that an enhancement N-channel transistor with
$V_{TN} > 0$, can become "ON" at $V_G = 0$ and work in the depletion mode. At the
contrary, the P-channel becomes more "OFF". For N-channel transistor, the
contribution of the interface states shows that the transfer curve can come
back toward its initial position.

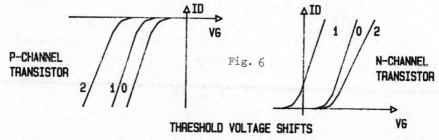

Fig. 6

THRESHOLD VOLTAGE SHIFTS

An other effect produced by the interface states is the decrease of the carrier mobility which is depicted by a decrease of the transfer curve slope.

As in bipolar transistors, leakage currents can be induced at the interfaces and into the field oxides.

At the technology level, the fabrication of MOS transistors with a good resistance under radiation is bound to the oxidation process and post-oxidation annealing treatment. Many studies have been carried out on: dry oxygen $SiO_2$, wet oxygen grown $SiO_2$, steam grown $SiO_2$, anodically grown $SiO_2$, deposited $SiO_2$ glasses and silicon-nitrogen compound ($Si_3N_4$)...Each manufacturer has his own process but the experiments concerning the relation of structure, processing and radiation effects have shown that the very clean oxides with regard to certain contaminants and having, as result low initial interface states and oxide charges are the best under radiation.

The reduction of the gate oxide thickness also will improve the radiation resistance.

It is to note that the deposit method of aluminium is important in the hardening process. Electron beam deposition can irradiate a MOS structure up to $1.10^4$ Gy(Si) and create defects which are annealed during the post-oxidation thermal process but which become again quickly apparent during a further irradiation. RF heating or crucible deposition methods will be preferred.

## 4.2. Dose rate effects.

The main effects induced by an ionizing radiation pulse are well described from the behaviour of irradiated reverse biased PN junction.

The figure 7 shows a cross section of a PN junction submitted to a radiation pulse. Electrons and holes are generated homogeneously into the semiconductor.

Under influence of high electric field in the depletion zone and of the diffusion process, associated to the high concentration variations (doping profiles) near the junction, the charges are separated and flow across the junction, giving rise to a photocurrent. [8]

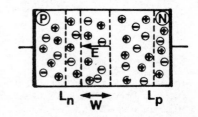

This photocurrent $I_P$ is described by the relation

$$I_P(t) = q.g_o.V_c.\dot{D}(t)$$

Fig.7

where $q = 1.6 \ 10^{-19}$ C
$g_o = 4.10^{15}$ carriers / $cm^3$. Gy(Si)
$g_o$ is the carrier generation rate conversion factor
$\dot{D}$ is the dose rate in Gy(Si)/s.
$V_c$ is the effective collection volume from which excess carriers can drift or diffuse to the junction before recombining.

To simplify, $V_c = A ( W + L_P + L_N )$ where A is the junction area, W the depletion region width and $L_P$ and $L_N$ the diffusion lengths in the N and P regions.

This expression of the photocurrent, $I_p = q g_0 V_c D (t)$, is often used to calculate the photocurrents of more complex structures.

In the case of bipolar transistors, primary photocurrents are generated within the depletion regions of the collector and emitter as well as within a few diffusion lengths of the junction in the base-emitter and the base-collector regions. Since the base-collector region is the largest in volume and backbiased, most of the primary photocurrent originates into the collector.

The photocurrent response of a transistor depends upon its application. The simple case to consider is the open base grounded emitter NPN transistor with a positive voltage to the collector. The primary base collector photocurrent makes the base region more positive and forward biases the emitter-base junction. The primary photocurrent acts as a base current which gives rise to a secondary photocurrent by the way of transistor action and the $h_{FE}$ current gain :

$$I_{SP} = I_{PP} ( 1 + h_{FE} )$$

In integrated circuits all of the active and parasitic junctions are the seat of photocurrents. Their contribution is more difficult to evaluate. (Parasitic junctions are the insulation junction of the integrated circuit monolitic structure).

It would be possible to take into account of conductivity changes induced by radiation in dielectrics and insulators; but with a few exceptions (as it will be seen farther with SOS) their effects remain weak compared to junction photocurrents.

In conclusion, the design of components with good resistance to the dose rate ionizing radiation involves:
- the reduction of dimensions of elementary diodes and transistors
- the reduction of carrier lifetimes into base and collector regions (gold doping)
- the use of appropriated isolation techniques of active elements in integrated circuits.

## 5. Vulnerability of integrated circuit technologies.

Among bipolar technologies, TTL (Transistor Transistor Logic) is the most widely used. Various series allow to choose circuits in function of speed or power consumption. Improvements are constantly brought to maintain them in the competition (Advanced Shottky TTL).

ECL (Emitter Coupled Logic) is the fastest technology but its high power consumption limits its use in embarked systems.

Last born one of bipolar logic, $I^2L$ (Integrated Injection Logic) is full of promises : high integration density and very low power consumption.

In MOS structures, two great technologies are sharing the market of logic and memory circuits : N-MOS and C-MOS. Their evolution trends to the dimension reduction as well as in the channel length as the oxide thickness (VLSI or VHSIC).

All of these technologies have been made the subject of many vulnerability studies under radiation.  It is not the purpose of this paper to give exhaustive results that it will be possible to find from other sources. But the TTL and C-MOS technologies are sufficiently representative of the two integrated circuit technologies to serve as examples in order to explain their behaviour under radiation and to suggest the possible improvements which can be made.

## 5.1. T T L .

The TTL basic structure is a three in-put NAND gate utilizing NPN transistors one diode and diffused resistors. (figure 8 )

### 5.1.1. Neutron effects.

Neutrons reduce the transistor current gains and increase the transistor $V_{CE}(SAT)$. These effects lead to a reduction of the output current.
The figure 9 displays the output current versus output voltage of a lot of NAND gates for two neutron fluences.

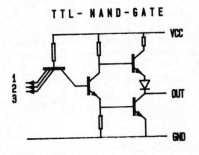

**TTL - NAND - GATE**

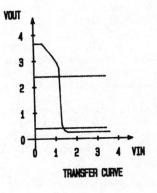

TRANSFER CURVE

Fig. 8

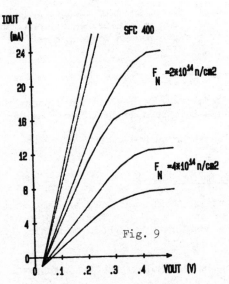

Fig. 9

The vulnerability threshold depends upon the TTL series (Low power , Normal, Schottky ...). It is between $5.10^{13}$ and $5.10^{14}$ n/cm$^2$ .

### 5.1.2. Ionizing dose effects.
TTL is not strongly affected by total dose. If the protection layer is of good quality, the electric performance characteristics only begin to change after $1.10^5$ Gy(Si).

### 5.1.3. Ionizing dose rate effects.

On the contrary, ionizing dose rates induce electric parasites into logic circuits the amplitude of which are susceptible to modify the logic states.

The radiation induced mechanisms are complex and merit analysis.

Standard bipolar integrated circuits are diffused in N-type epitaxial silicon layer. Active elements, transistors and diodes are insulated between them by reverse biased PN junctions obtained by diffusion of P regions into the N epi-layer.

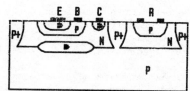

Fig. 10

Resistors also are diffused and junction insulated.

So, to each NPN transistor is associated a parasitic PNP transistor and each resistor is shunted by a distributed junction. ( figure 10 )

Under a radiation pulse, photocurrents are generated into all junctions and transistors. But due to the larger area of insulating junctions, their photocurrents are more important and contribute to a relatively modest vulnerability threshold (logic upset)   : $2.10^6$ to $2.10^7$ Gy(Si)/s.

At high dose rate levels, these photocurrents are susceptible to burn-out the junctions and the metallizations.

An other mechanism can be induced into the IC structure and cause catastrophic failure. Known as Latch up, it takes place in   four layer structures (NPNP or PNPN) where photocurrents and applied voltages can give rise to thyristor effects, inducing anomalous logic states and extra sustained supply currents.

This effect is difficult to get in TTL but it exists and depends on the integrated circuit design.

In order to become to latch up free, a solution leads to use a dielectric isolation technique of active elements.

The basic technology process (EPIC) Epitaxial Process for Integrated Circuits, is more complex than classical process. It consists to isolate small monocrystalline silicon moats into a polycrystalline silicon layer by silicon oxide layers. The poly-silicon layer (300 µm thick) is used as a mechanical support. Active elements and resistors are diffused into the monocrystalline regions. (figure 11)

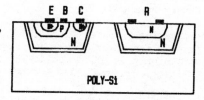

Fig. 11

Other techniques can be associated to the dielectric isolation in order to improve the behaviour of TTL under transient radiation :

> - the substitution of diffused resistors by
>   vacuum deposited resistors ( NiCr, SiCr,... )
>
> - gold doping, to decrease the base and
>   collector minority carrier lifetimes
>
> - the integration of small geometry transistors
>   with high $F_T$ .

The use of these techniques has allowed to improve the transient vulnerability threshold by at least two orders of magnitude ($2.10^8$ Gy(Si)/s) and to eliminate the latch up phenomenon. Other improvements can be available with an optimization of the circuit design.

These technologies have been implemented by many manufacturers particularly by Harris, Texas, Motorola, RCA in USA, Toshiba in JAPAN and Thomson in FRANCE.

In an other field, they are really suited to manufacture high voltage or linear circuits with complementary bipolar transistors.(figure 12)

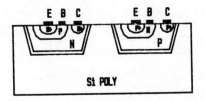

Fig. 12

Remark. This dielectric isolation technique would be advantageously used for $I^2L$ structures. It could improve the PNP lateral transistor which is the weak point of $I^2L$ under neutrons.

## 5.2. C - MOS.

The basic C-MOS cell is an inverter including two complementary transistors, N-and P-channel, both working in the enhancement mode. [10]

The inverter transfer curve and supply current observation shows the power consumption only is effective during the transition. (figure 13)

This structure is very well adapted to high integration density.

### 5.2.1. Neutron effects.

Because the MOS transistors are majority carrier devices, C-MOS are not too much affected by neutrons. Induced ionizing effects have to be taken into account with ionizing total dose.

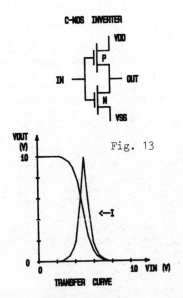

Fig. 13

### 5.2.2. Ionizing total dose effects.

The predominant effect of ionizing radiation is the threshold voltage shifts of **both** N- and P- channel transistors. As a result, the transfer curve is shifted towards the left. [11]

In addition, surface effects are induced into the field oxides giving rise to leakage currents into parasitic MOS transistors which can disturb the circuit behaviour.

The vulnerability level is strongly dependent on fabrication conditions and circuit type. It will be determined by the maximum variations of the tolerable threshold voltage shifts. So, the total dose limit for commercial microprocessor is not more than 100 Gy(Si); but for simpler logic circuits it can be about 1000 Gy(Si).

In using appropriate processes to manufacture ultra clean gate and field oxides, it is possible to improve the radiation tolerant level up to $10^4$ Gy(Si). (example: Rad-hard CD 4000 series with Z suffix RCA )

### 5.2.3. Ionizing dose rate effects.

As in the classical bipolar monolithic structure, photocurrents are generated in all of the junctions. The first radiation investigations connected with C-MOS inverter have been carried out by A.G. Holmes-Siedle and al. in 1967 but they are always of actuality. [12]

Three photocurrents have to be considered as shown in the figure 14 $I_W$ is the photocurrent generated into the P-well junction, where is diffused the N-channel transistor. This one can be important at high radiation rates but flows from supply to ground increasing by this way the supply current. The two other photocurrents to take into account are $I_P$ and $I_N$ created into the drain junctions of P-and N-channel transistors.

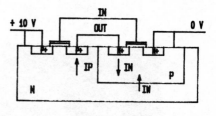

Fig. 14

They give rise to an induced voltage at the inverter output :

$$\left| \Delta V \right| = R_{P(N)} \cdot \left| I_P - I_N \right|$$

where $R_{P(N)}$ is the ON resistance of the transistor in conduction state. This equation shows a certain photocurrent compensation due to the complementarity of the transistors.

The small transistor dimensions and this compensation effect would have induced a high dose rate tolerant level. But the radiation analysis have shown that this level was much lower ( $\simeq 1.10^6$ Gy(Si)/s ) than this one obtained by the contribution of known photocurrents.

The reason is similar to this one found with the bipolar structure. Latch up takes place in four layer structures giving rise to anomalous photocurrents. Here the effects are more important and can be easily demonstrated. They can lead to burn-out junctions, metallizations and bonding wires ( photo) Many radiation studies and experiments have been implemented . [13]

LATCH-UP  INDUCED  EFFECTS

Metallization burning-out

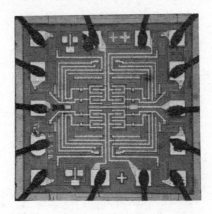

Bonding wire burning-out

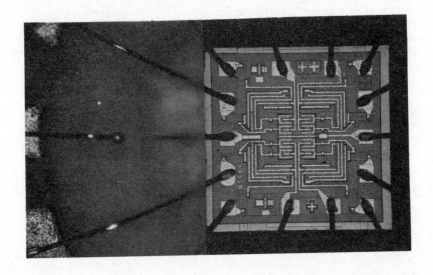

## 6. Latch-up free C-MOS technologies.

### 6.1. Isolation technology.

According to the foregoing, the transient radiation failures  are the result of the active element insulation mode into the integrated circuit monolithic structure.

It would be possible to manufacture C-MOS transistors in using dielectric isolation as for TTL. Harris used this technique for CD 4000 series in '73, but it was not suited to high integration level.

Other techniques have been developped which consist of local oxide isolation as is shown in the figure 16

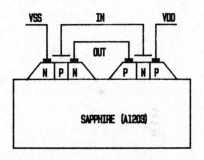

Various names are given to those   : Isoplanar, ISO-CMOS, LOC MOS ... They are the subject of transient radiation investigations. But it is difficult and even impossible to obtain a final proof to a 100 % certainty level to have free latch-up circuits.

Fig. 16

### 6.2 Al-gate C-MOS/SOS technology.

The Silicon On Sapphire (SOS) technology allows the total insulation of active elements without semiconductor junctions. It leads to improve the circuit speed in reducing the parasitic capacitances, to increase the integration density and to increase the transient radiation level.
The SOS technology main point lies  in the use of monocrystalline alumina substrate ($\simeq$ 300 µm thick) with an oriented plane ( $1\bar{1}02$ ) on the top of which a thin silicon layer (0.6 µm) has been epitaxied with (100) orientation.

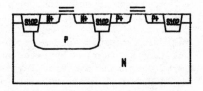

Fig. 17

The first steps of the processing consists of the creation of silicon islands where the transistors are diffused or implanted. The gate oxide is generally thermally grown and aluminium is vapor deposited upon the oxide.

The figure 17  shows the C-MOS/SOS inverter structure.

This technology has been implemented by RCA, Rockwell, Hewlett-Packard and Hughes in USA, by Siemens in Germany and by Thomson-Efcis in France.

### 6.2.1. C-MOS/SOS radiation vulnerability.

### 6.2.1.1. Ionizing total dose effects.

To the creation of positive charge into the gate oxide and of the electronic states at the Si-SiO$_2$ interface, it is necessary to add the same effects created into the sapphire substrate and at the silicon-sapphire interface.

The two first effects give rise to threshold voltage shifts leading to the shift of transfer curve. In addition, leakage currents appear. They are due to the presence of parasitic lateral transistors onto the edges of silicon islands. (figure 18)

The radiation positive charge created into the sapphire substrate can induce an electron inversion layer in the silicon near the silicon-sapphire interface, connecting source to drain and giving rise to the so called back channel leakage current. [14]

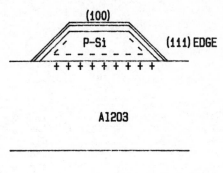

Fig. 18

This current is in general independent of the gate voltage and hence cannot be turned off. Some ion implantation processes are performed to decrease this effect and to allow a good ionizing total dose level ($1.10^4$ Gy(Si))

### 6.2.1.2. Ionizing dose rate effects.

Transient effects into C-MOS/SOS circuits are the result of the active junction photocurrents and of the induced photocurrents by the sapphire photoconduction.

Intrinsic junction photocurrents are very small, due to the small junction areas. With the considered silicon layer thickness, the photocurrent value is about $3.10^{-15}$ A/Gy(Si) per micron of length.

But all of the silicon stripes and aluminium interconnexions are biased at various voltages which induce semicircular electric field lines into the underlying sapphire. Under the influence of these electric fields, the electrons generated during the ionizing pulse are collected and act as photocurrents (figure 19)

The involved mechanisms have been described by D.H.Phillips. From this analysis, it results that the sapphire induced photocurrents are 2 to 5 times more important than those issued of the silicon structure. [15]

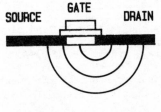

Fig. 19

The logic upset threshold of C-MOS/SOS integrated circuits is found between $5.10^8$ and $1.10^9$ Gy(Si)/s, showing an improvement of about two orders of magnitude with regard to classical C-MOS circuits; of course, the C-MOS/SOS circuits are latch up free.

6.3. <u>Si-gate C-MOS/SOS technology</u>.

As for all of the MOS technologies, the Si-gate process tends to replace the Al-gate process. That contributes to improve the electric performances and allow the manufacturing of more complex circuits by the way of a dimension reduction (channel length and oxide thickness) inducing a higher integration level (VLSI and VHSIC).

Actually, microprocessor or memory circuits are already on the market (GPU 001, MWS 5114 ...RCA). Custom and semi-custom integrated circuits are becoming available (RCA, HAFO, Marconi ...).

The previous circuits work with the two transistors in the inhancement mode. Another working mode where the N-channel is off when it is deep depleted leads to easier technology (with less masks) which is in processing at Thomson-Efcis and in radiation evaluation at CEA.

Good results already have been obtained under high radiation total dose levels (> $10^5$ Gy(Si)). Transient effects are about the same than those found with the Al-gate C-MOS/SOS technology.

7. <u>New trends in radiation tolerant technologies</u>.

7.1. <u>SOI (Silicon On Insulator)</u>.

One of the reasons of the want of success of the SOS integrated circuits on the commercial market is the cost of the substrate which requires a lot of preparing work and characterization. But, the techniques used with SOS to improve the electric properties of silicon (recrystallization by laser annealing or others) have lead to reconsider the substrate problem.

SOI technology consists of growing silicon layers, on the top of silicon oxide layers, which will be recrystallized by various appropriate methods (laser annealing, liquid and/or phase epitaxy, strip heater ...). Those latter are still at the research level and have to demonstrate their ability to be industrial. It is too soon to give an opinion on SOI technology but the whole of the technology processes involved, notably to manufacture 3 D devices, require our attention.

As regards to the under radiation vulnerability of this technology, it is also to soon to pronounce one's opinion.

Many insulating layers and many interfaces will be the seat of well known ionizing effects. But, the hardening techniques used with the SOS technology, which is a particular case of SOI technology will can be implemented without difficulty.

7.2. <u>MNOS. Metal Nitride Oxide Semiconductor</u>.

Generally, the hardening levels of memory devices are low . N-MOS dynamic memories are disturbed after 10 Gy(Si) of ionizing radiation and their logic level at $10^6$ Gy(Si)/s.

C-MOS and C-MOS/SOS static memories are better : $10^2$ - $10^4$ Gy(Si) for prolonged ionization radiation and $2.10^8$ Gy(Si)/s for the logic upset (CMOS/SOS).

For specific applications it would be interesting to have a "harder" memory component.

The MNOS, which is a metal-insulator-semiconductor transistor with two insulators (SiO$_2$ layer $\simeq$ 30 Å thick and Si$_3$N$_4$ layer $\simeq$ 500 Å thick) is a non-volatile device. Integral part of a memory cell, this transistor can keep its information up to a memory upset of $10^{10}$ Gy(Si)/s.

The MNOS technology is compatible with other MOS technologies. Associated to SOS Technology this one would constitute the best radiation tolerant device for Integrated Circuits.

Actually, many manufacturers are involved with C-MOS, M-NOS/SOS technology : Rockwell, Westinghouse [16]. Sandia [17] develops hardened MNOS memory devices associated to a bulk C-MOS technology. Plessey and Phillips have also devices on the commercial market.

## 7.3. GaAs (Gallium Arsenide).

In the field of the electronic technologies for the future, Gallium Arsenide (GaAs) is well placed.

Already, GaAs field effect transistors have demonstrated their ability to work into power and frequency ranges not accessible to silicon.

As for integrated circuits, GaAs technology is at the beginning of its development and many research laboratories are working on the subject.

In a near future, it would be very interesting to concept electronic systems associating GaAs components such as optoelectronic devices, GaAs linear and digital integrated circuits and GaAs power components.

Ga As Integrated circuits are still under development ; the information concerning radiation effects on GaAs is less voluminous than for the silicon.

However, some characteristics of the technology and the first investigations made on material and devices tend to demonstrate that it is well adapted to manufacture radiation tolerant integrated circuits :
First point, the semi-insulating nature of gallium arsenide constitutes a good material characteristic for a substrate to avoid latch up
Second point, the use of GaAs field effect transistors with PN-junction or schottky barrier gate structure which are not too much susceptible to total dose of ionizing radiation.

Under radiation, GaAs material and devices undergo the same degradation types than the silicon case.

Displacement damages take place into the crystal lattice where are found the two defects : creation of vacancy interstitial pairs damaged regions which are different of those created in silicon (spike zones of quasi-metallic behaviour instead of defect clusters in silicon.) They induce similar effects on carrier lifetimes, carrier removal and mobility degradation. But, due to the fact of carrier lifetimes are very short in GaAs the degradations are relatively less important compared to silicon.

The basic element of GaAs integrated circuits is the MESFET (Metal Semiconductor Field Effect Transistor). There is no active oxide into the structure so that under ionizing radiation the behaviour of GaAs is comparable to the bipolar technology.

R. ZULEEG and K. LEHOVEC [18-19] have shown some results on GaAs JFET where neutron fluence of $1,7 \; 10^{15}$ n/cm$^2$ and ionizing total dose of $1.10^6$Gy(GaAs) do not      modify    strongly the intrinsic characteristics. The same performances are expected for integrated circuits.

Transient effects are also similar to those in bipolar circuits. Photocurrents generated in   PN-junction   can be approximated by :

$$I_P = q \; go \; Vc \; \dot{D} = q \; go \; ALc \; \dot{D}$$

where  A is junction area

Lc is an effective collection length

go here  is the carrier generation constant factor for gallium arsenide. Its value is $6,63.10^{15}$ carriers/cm$^3$.Gy(GaAs). It is 50% larger than that for silicon.

The first results give a dose rate for logic upset in the range of $10^8$ to $1.10^9$ Gy(GaAs)/s.

Because of the good behaviour under ionizing radiation and the minimum effects of neutrons, GaAs is a fully promising technology for devices and intregrated circuits able to work in severe nuclear environments.

## 8. Conclusion.

The radiation environment induces in material and devices severe degradations which can be reduced and even eliminated with hardening methods. Those latter are very often improvements of the basic technology. Neutrons and ionizing radiation (total dose) defects are often of most concern in fabrication processes which have  to tend to the best quality.

In integrated circuits, the more important defects (latch up) are due to the monolithic structure which in classical technologies consists of a junction isolation of active elements. The dielectric isolation with polycrystalline silicon and the silicon on sapphire technologies have notably improved the behaviour of logic components under radiation pulses.

Actually, the C-MOS/SOS Si-Gate technology is the best one to manufacture complex integrated circuits with a good resistance  to the radiation effects.

## 9. Acknowledgements.

The author would like to thank M. ROUMEGUERE and A. BURKHART for many helpful discussions.

## 10. References.

[1] "Theoretical and Experimental determination of neutron energy deposition in silicon"
E.C. SMITH and All.
IEEE Trans. on NS - December 1966 - pp 11-17

[2] "Energy dependence of neutron damage in silicon"
H.J. STEIN
J. of A.P. - January 1967 - pp. 204 - 210

[3] Thesis n° 1736 - Université Paul Sabatier Toulouse
J. SAURET - Mai 1975

[4] "Energy dependence of displacement effects in semiconductors"
V.A.J. Van LINT   R.E. LEADON
IEEE Trans. on NS - December 1972 - pp. 181-185

[5] "Design curves for predicting fast neutron - induced resistivity changes
in silicon"
M.G. BUELHER
Proceedings of the IEEE - October 1968, pp. 1741 - 1743

[6] A survey of radiation effects in metal - insulator - semiconductor
devices
K.H. ZAININGER and A.G. HOLMES SIEDLE
RCA Review vol. 28 n° 2 June 1967

[7] Effects of ionizing radiation on oxidized silicon surface and planar
devices
E.H. SNOW, A.S. GROVE, D.J. FITZGERALD
Proc. of IEEE - vol. 58 n° 7 July 1967 - pp. 1168 - 1185

[8] Radiation effects in semiconductor  devices
F. LARIN
Wiley

[9] Hardness of MOS and bipolar integrated circuits
D.M. LONG
IEEE Trans. on NS - vol. 27 n° 6 - Dec. 1980

[10]Switching response of complementary - symmetry MOS transistor logic
circuits
J.R. BURNS
RCA Review Dec. 64

[11]Permanent radiation effects in complementary - symmetry MOS integrated
circuits
W. POCH and A.G. HOLMES SIEDLE
IEEE Trans. on NS - Dec. 69

[12]TREE. Effects in complementary symmetry MOS transistor integrated
circuits
W.J. DENNEHY - A.G. HOLMES SIEDLE
IEEE Trans. on NS - Dec. 69

[13]Latch up in CMOS Integrated circuits
B.L. GREGORY and B.D. SHAFER
IEEE Trans. on Nuclear Science Dec. 73

[14]Radiation induced leakage currents in silicon on sapphire MOS transistors
S.T. WANG and B.S.H. ROYCE
IEEE Trans. on NS - vol. NS.23 n° 6 - Dec. 76

[15]Complementary - Symmetry/ Metal oxide semiconductor (C-MOS)
Circuit hardening
D.H. PHILLIPS - R.K. PANCHOLY
AFWL - TR 74-243 - vol. 2 - June 75

[16] Hardened MNOS/SOS Electrically reprogrammable non volatile memory
J.R. CRICCHI, M.D. FITZPATRICK, F.C. BLAHA and B.T. AHLPORT
IEEE Trans. on NS - Dec. 77

[17] A Radiation hardened non volatile MNOS RAM
T.F. WROBEL, W.H. DODSON, G.L HASH, R.V. JONES, R.D. NASBY
SANDIA NATIONAL LAB. AND OLSON
AIRFORCE WEAPONS LAB.
IEEE 20th conf. on Nuclear and Space radiation effects July 18-20 1983

[18] Radiation effects in GaAs junction field effect transistors
R. ZULEEG and K. LEHOVEC
IEEE Trans. on NS vol. 27 oct. 80

[19] Ionizing radiation response of GaAs JFETS and DCFL circuits
R. ZULEEG and J.H. NOTHOFF and G.L. TROEGER
IEEE Trans. on NS 29 - Dec. 82

*Inst. Phys. Conf. Ser. No. 69*
*Paper presented at ESSDERC/SSSDT 1983, Canterbury 13–16 Sept. 1983*

105

# Device failure mechanisms in integrated circuits

J. Schädel, Siemens AG, Balanstr. 73, 8000 München 80

Abstract A survey is given for the most prominent failure
mechanisms of state-of-the-art ICs, followed
by a review of the single mechanisms and their
probable impact on future trends in IC development.

## 1. Introduction

As far as development and production techniques are concerned, ICs are microelectronics products that have been
well mastered. That this is so is evident from, among other
things, the fact that in the twenty years ICs have existed
the failure rate for an IC gate has dropped by a factor of
10,000, Fig.1. This
is largely due to
the dramatic reduct-
ion in the area of
silicon a gate
requires. Minia-
turization is
continuing and
reliability
figures for
state-of-the-
art VLSI circuits
are most satisfac-
tory these days –
a few hundred
failures in $10^9$
operating hours.

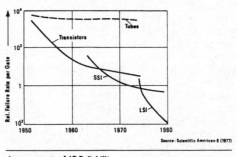

Improvements of IC Reliability

Fig. 1

Why, then, talk about failure mechanisms? There are at
least two good reasons:

1. IC users are not satisfied.
2. Some failure mechanisms that have
been controllable until now will
become a threat as a result of
further miniaturization.

IC users have in-
creased the number
of ICs in their
devices and
onboard consider-
ably. A necessary
corollary of this
was that IC failure
rates had to be
radically reduced
if the device
manufacturer's
yield and his
devices' field
failure rate
were to remain
economically
acceptable. (Figure 2)

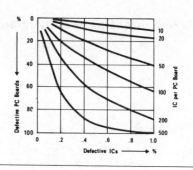

**PC Board Yield vs. Quality of ICs**

Fig. 2

Interest here centered on the early failures (infant
mortality); and it still does. Experience shows that
infant mortality failures are almost exclusively caused by manufacturing defects; and they are characterized by a rapidly
falling failure
rate in the first
few hours of
operation. With
regard to the
celebrated bath-
tub curve, the
high initial
failure rate
is caused by
infant mortality
before the curve
flattens out
(operating phase).
As infant mortality
failures reflect more
a learning curve and are

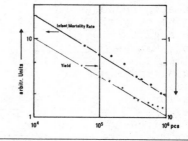

**IC Learning Curve**

Fig. 3

not largely determined by intrinsic defect mechanisms,
they should be demonstrated by just one figure. In
Figure 3, the bottom curve is a typical learning curve.
The failures during manufacture are given versus the total
number of components manufactured. The top, parallel curve
depicts the development of the products' infant mortality
rate. (The products dealt with here are MOS LSI devices.)
Fortunately it is possible, at the start of the learning
curve, to reduce infant mortality rate by implementing
screening measures such as burn-in or voltage stress.

Let us now examine which failures are of significance in
the subsequent flat section of the bath-tub curve, i.e. in
actual operation, and which parts of an IC are involved.

This frequency
distribution is
based on defect
analyses by
testing insti-
tutes, field
failure reports
from users,
reliability
reports from
nine well-known
manufacturers,
and our own
experience with
ICs from at least
20 different
suppliers.(Figure 4)

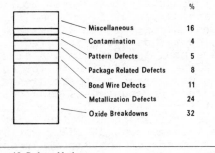

|  | % |
|---|---|
| Miscellaneous | 16 |
| Contamination | 4 |
| Pattern Defects | 5 |
| Package Related Defects | 8 |
| Bond Wire Defects | 11 |
| Metallization Defects | 24 |
| Oxide Breakdowns | 32 |

IC Defect Modes

Fig. 4

All types of ICs manufactured between about 1975 and 1980
have been added together: SSI through LSI, cavity and
plastic packages, MOS and bipolar ICs, CMOS and single-
channel circuits, commercial and consumer products.
This is, of course, only permissible if shared characteristics are to be considered and the products have a number
of important types of defect in common. Comparison of this
with a similar diagram for 1970 would reveal different defect ratios, because we would be dealing primarily with SSI
devices, but the same defect types. We have succeeded in
improving our control over failure mechanisms, as is illustrated by the excellent reliability figures for present-day
ICs, but the complete elimination of such mechanisms is
evidently a dream that has not come true.

These findings will be no surprise to the expert, who knows
that in the last 10 years the number of process steps required for a typical process has increased approximately by
a factor of 2. Typical line width has been reduced from about
12.5 µm to 2.5 µm, gate oxide thickness from 150 nm to 40
nm, and penetration depth of diffusion zones from 3 µm to
0.3 µm. We know that the trends described are continuing
unabated. It is all the more astonishing that there has been
only an insignificant change in the failure types.

Today, we shall be talking about old friends, most of whom
have been with us since ICs were invented. Expressions such
as "weak spots" (in oxide layers), "purple plague" or
"voids" (in metal lines) are almost as old as "IC" or
"planar technology".

Before discussing the mechanisms, we must delimit our subject
still further. We shall not be considering insignificant
manufacturing defects, even if they do have an important
effect on practical operations. Each of you will know that
a metal line that has been half scratched away already will

burn through at its weakest point at the first opportunity, and each of you will know how this problem can generally be avoided. Neither shall we be talking about systematic or coincidental development defects; these occur, of course, but are usually detected and eliminated during the development phase. In addition to insignificant manufacturing and development defects, there are a series of failure causes which must be counted as manufacturing defects in their widest sense but which cannot be completely mastered because not enough is known about them. Put another way: if we knew more about them, we might be able to master them by means of appropriate production controls. It is this type of failure mechanism that I should like to talk about.

Let us now take a look at where these special gremlins are hidden on the IC. On average, taking all types of ICs into account, chip defects make up the majority of defects (64%), followed by defects around the bonding wires (20%) and package defects (16%). These percentages vary greatly in practise, depending on the manufacturer, the technology, the degree of integration, the number of pins on the package and the type of package. There are, of course, various mechanisms that are the cause of the types of failure described: a short-circuit through an oxide layer may be a genuine dielectric breakdown or a time-dependent breakdown, or may be caused by a pin hole or a locally reduced breakdown voltage.

The types of failure to be discussed, which have been selected arbitrarily, are characterized by the fact that the failure mechanisms affect virtually all IC manufacturers, have a long history, cannot be eliminated completely because there are too many influencing factors (or too many unknown ones) and will probably play an important role in the future too.

2. Chip-related failures
Let us start with the chip. The following failure mechanisms can be observed:

- Current flow through dielectric isolation layers
- Transport action in isolation layers
      Ion drift
      Hot electron trapping
- Transport action in metal layers
      Void/hillock formation, electromigration
      Corrosion
- Transport action in barrier layers
      Phase formation in contact holes
      Phase formation at wire bonds

2.1 Oxides

Oxide defects, particularly with regard to gate oxides, are of prime importance. If MOS capacitors with dielectrics in the form of thin oxide layers are investigated by applying a voltage which is allowed to rise slowly to a field strength of around 10 MV/cm, varying breakdown behavior patterns are recognizable as a rule. The capacitors can be categorized into various types. These are:(Figure 5)

%
40
30
20
10
Intrinsic Breakdown
Shorts
0
$V_{max}$

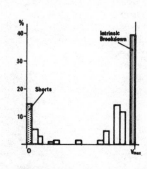

**Gate Oxide Breakdown Voltages**

Fig. 5

| | |
|---|---|
| Type A | capacitors with a breakdown voltage of 0 V; |
| Type B | capacitors whose breakdown voltages are markedly below the theoretical value in line with $10^7$ V/cm; |
| Type C | capacitors whose breakdown voltages rise or fall because of the influence of the electric field; |
| Type D: | capacitors with an intrinsic breakdown voltage. |

Type A defects can be attributed simply to holes in the dielectric. Type B and C defects have many things in common (Figure 6). Type B defects are characterized as follows /1/2/3/4/: The defects are restricted to certain areas (diameters 0.1 to 2 µm) and are distributed statistically. The temperature dependence of the breakdown voltage is greater than for the intrinsic breakdown through

TDDB
Thermal Breakdown
Field Breakdown
local spots
thickness dependent
caused by contamination
and/or imperfections
generating local current paths via enhanced conductivity, thermal damage by Joule heating
generating trapping centers for hot electrons, impact ionisation by locally enhanced field strength

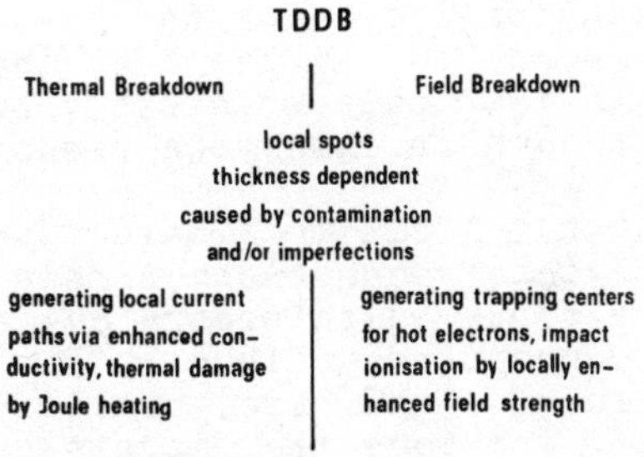

**Time Dependent Dielectric Breakdown**

Fig. 6

impact ionization. The defect density per $cm^2$ can
typically be reduced by using particle-free chemicals
during processing or by annealing the silicon wafers before
actual oxidation.
Sometimes defect
density depends
heavily on the
oxide thickness.
These are almost
certain to be
thermal break-
downs. Their
"breakdown volt-
age" is deter-
mined by the
rival influences
of the Joule
heat generated
by a current
through the

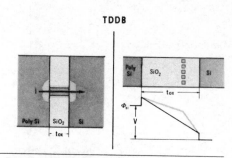

**TDDB**

**Time Dependent Dielectric Breakdown**

Fig. 7

oxide and of the defect's immediate vicinity which controls
the heat dissipation. The actual breakdown is characterized
by thermal destruction of the dielectric when the current
becomes so great that the heat generated is no longer
dissipated, Figure 7.

Type C breakdowns are also time-dependent dielectric breakdowns. Their behavior under the influence of fields greater
than $3 \times 10^6$ V/cm is such that their breakdown voltages
change either in fractions of a second or in thousands of
hours, i.e. they usually drop and, much less often, may also
rise /5/6/. Such breakdowns are an undesirable reliability
risk, because failure rates on account of this mechanism
may change by more than 30 orders of magnitude between 5
and 10 MV/cm of field strength applied to the oxide. And
unfortunately this time behavior is heavily dependent on the
oxidation temperature, the oxidation conditions and, possibly, the perfection of the silicon substrate.

The high degree of dependence on the size of the area under
investigation indicates the presence of dot-like defects.This
is also borne out by the fact that the breakdown behavior
can be influenced considerably by adding HCl to the oxidation
medium though it cannot be rendered totally time-independent.

No one can yet claim to have found a satisfactory explanation for this phenomenon. A selection of models is given in
/7/. The observed phenomena all occur in the region of the
tunnel currents described by Fowler and Nordheim. It is therefore assumed - and here, apart from slight variations of detail, all previous test models concur - that the high field
strengths cause hot charge carriers to tunnel through the
oxide and become trapped /8/. Charged traps of this kind
increase the internal field strength in the oxide locally
up to the intrinsic breakdown point.

The time required for this is determined by the density of the tunneling electrons, the density and the capture cross section of the    unoccupied traps, and also by any de- trapping which might take place. The higher the externally applied field, the lower the charging required to reach the breakdown point.

So much for the presentation of the models, which illustrate the observations rather well, but don't tell us too much about the origin of the defects.

Thermal breakdowns (type B) and time-dependent dielectric breakdowns (type C) are evidently linked with local "oxide defects". Manufacturing facilities are becoming cleaner and cleaner, purification methods more and more effective.  However, the demands made in particular on the blocking proper- ties of thin dielectric layers and their defect density are becoming still higher too. We must conduct more fundamental research and thus do more to provide an explanation of the nature of oxide defects. That this is necessary is also borne out by the fact that, more and more frequently in the course of further miniaturization, we must reckon  with the presence of hot charge carriers which can easily penetrate oxides (and not only gate oxides) and become trapped there, destabilizing the transistors' electrical properties. A good example of the fact that it is not just gate oxides that matter is provided in /9/: in a very narrow, short-channel transistor, the high-trap-density birds-beak area is apparent from the presence of a gate-voltage-dependent gate current. Charge trapping in oxides will be one of the most significant failure mechanisms in the near future, even if trapping does not cause oxide breakdown straightaway.

2.2 Electromigration
The future signi-
ficance of another
failure mechanism,
electromigration,
is considered
similar to that
of hot electron
trapping. Electro-
migration pri-
marily affects
the aluminum
lines: under the
influence of
current densities
greater than
$2 \times 10^5$ A/cm$^2$
material transport
takes place from anode to cathode, as does electron con-

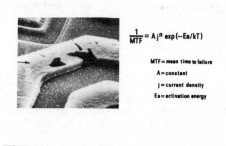

$$\frac{1}{MTF} = A j^n \exp(-Ea/kT)$$

MTF = mean time to failure
A = constant
j = current density
Ea = activation energy

Electromigration

Fig. 8

duction (in the opposite direction). If this material transport action does not take place uniformly over the conductor
section, it may cause conductor paths to be cut through
void formation after a few hours; Figure 8.

This type of defect does not play a dominant role in our
defect mode distribution: occurrence was usually due to
manufacturing defects (weak points were the aluminium
lines over oxide steps, aluminium line constriction etc.).
The equation governing the phenomenon is straightforward:

$$\frac{1}{MTF} = Aj^n \exp(-E_A/KT)$$

MTF = Mean time to failure
A   = Constant
j   = Current density $(A/cm^2)$
$E_A$ = Activation energy
K   = Boltzmann's constant
T   = Absolute temperature

By adjusting A,
n and $E_A$, this
equation permits
a good description
of the results
of the investi-
gation; but it is
not suitable for
forecasting life-
times.Rarely will
an effect be sub-
jected to as many
influencing
factors as
electromigration.
Undoubtedly, this
is partly due to
the fact that we are actually dealing with a group of
effects, see Figure 9.

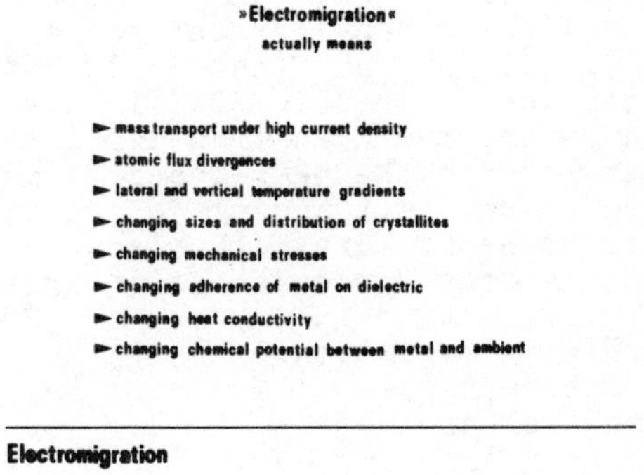

»Electromigration«
actually means

► mass transport under high current density
► atomic flux divergences
► lateral and vertical temperature gradients
► changing sizes and distribution of crystallites
► changing mechanical stresses
► changing adherence of metal on dielectric
► changing heat conductivity
► changing chemical potential between metal and ambient

**Electromigration**

Fig. 9

Only the first effect of Figure 9 deserves the title
"electromigration". The high current densities it requires
produce thermal effects, of course: a current of
$10^4$ A/cm$^2$ will melt a suspended aluminum wire.

At a current level of 4 mA, the 2 µm aluminum lines
on  today's ICs already exhibit current densities of $10^5$ A/cm$^2$.
This is why solutions are being sought, throughout the world,
as to how this latent reliability risk can be reduced.

If material transport were uniform along and across the
conductor path, acceptable lifetimes could be attained using

pure-aluminum paths. Since, however, transport takes place largely by means of grain boundary diffusion, there are always locations where the material flow branches off and leaves behind cavities which themselves cause a local increase in current density and thus enhance the action. Long conductor paths are particularly susceptible to this: the longer the path, the more likely such an action will take place. Narrow conductor paths in particular are at risk: it is fairly likely that these actions will cause not only voids but also fractures. In very narrow conductor lines, in which the grain diameters are larger than the path width, an increase in lifetime can be observed, doubtlessly because grain boundary diffusion parallel to the direction of current can no longer take place /10/11/.

An increase in lifetime can also be attained by introducing other metals as additives in order to suppress grain boundary diffusion. (Copper is particularly effective in this regard.) The search for other additives is well under way because copper, i.e. its $Al_2Cu$ alloy, occasionally precipitates and is also suspected of causing grain boundary corrosion. Nickel, chromium and magnesium have already been recommended; but they too have their drawbacks, though different from the disadvantages of copper.

At present, it is particularly fashionable to recommend producing fine-grain conductor paths with low crystallite-size dispersion. It is indeed true that, because many diffusion paths can be provided in parallel, an increase in lifetime can be attained. However, there are many influencing factors in the manufacture and operation of ICs that cause the aluminum to be recrystallized into larger crystals with less free energy.

Passivating the aluminum lines also helps to increase lifetime, though it is not entirely clear to what extent this is primarily due to an improvement in heat dissipation or to changes in surface condition. There are also environmental influences. Investigations in $H_2$, $O_2$, $H_2O$ and Ar atmospheres have shown how complex the events are. Both increases and decreases in lifetime were reported for one and the same gas /12/13/.

In March 1983 at the International Reliability Physics Symposium in Phoenix, Arizona, Thomas and Calabrese /14/ of Rome Air Development Center presented a film in which electromigration in aluminum conductor paths had been observed using time lapse techniques. This film showed a number of remarkable results that challenge at least some of the suppositions hitherto made about electromigration phenomena: Voids migrated through the conductor path without being affected by grain boundaries. In the midst of fairly large voids, islands formed and disappeared. The conductor path was enveloped by a thin, elastic film ($Al_2O_3$?) which was not pierced by the hillocks that

were forming. As the voids migrated further, the old
surface structure was restored; voids migrated both
counter and perpendicular to the electron flow. Thus
there is sufficient scope for investigation; and, without
doubt, our prospects will be based on new discoveries.

## 2.3 Contact holes
For the last of the chip-related reliability problems, let
me now mention the problem of contact holes. Contact
holes have always tended to fail, either because of ex-
cessive contact resistance or because of overalloying:
We have learned to live with overalloying. Aluminum can
contain up to 1.6% silicon in solution. When pure aluminum
comes into contact with silicon, a specific quantity of
silicon solubilizes at temperatures between 400 and $550^{\circ}$C.
This has no effect, provided the contact hole area dissolves
uniformly, which is but difficult to achieve. This dissolv-
ing can be avoided almost entirely if approximately 1-2%
silicon is added to the aluminum beforehand. In this way,
the problem of "spiking" is solved.

The smaller the contact size, the more often the cause of
failure is a substrate short-circuit in the contact hole.
Gargini, Tseng and Woods /15/ have pointed out that the
current density in contact holes increases by $k^3$, where k
is the scaling factor, if the voltage supply in the chip
remains constant:

$$j \sim k^3$$

But that's not all: at the edge of the contact hole, there
is a further increase in current density - current crowding
- in the direction of current flow.
The high current
density at the
front edge of the
contact hole
(Figure 10) can
cause silicon
migration; this
in turn hollows
out the contact
hole down to the
PN boundary
below and thus
produces a sub-
strate short-circuit.
The average life-
time of a contact
hole of this type

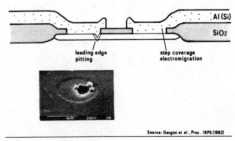

Contact Holes

Source: Gargini et al., Proc. IRPS (1982)

Fig. 10

is roughly proportional to the junction depth. Consequently
the application of a so called "plug-in-diffusion", which
enhances junction depth under the contact hole helps greatly
to improve lifetimes. To avoid any electromigration of silicon

from the contact
hole to the alu-
minum, Gargini
et al. recommend
the use of a
barrier metal
layer between
aluminum and
silicon.They
particularly used
tungsten (Figure 11).

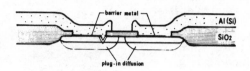

**Contact Holes**

Source: Gargini et al., Proc. IRPS (1982)

Silicon migra-
tion from Al(Si)
can segregate
p-doped silicon at
the reverse-polarity
contact hole, increasing   the contact resistance.

Fig. 11

The third electromigration-based failure mechanism is a
small contact hole which, surrounded by a relatively high
oxide step, can completely insulate itself through aluminum
migration from the oxide edge.

All three failure mechanisms will tend to become more significant as the contact hole area is reduced.  The risks will
multiply, too: as the contact hole becomes smaller, oxide
etching in the contact hole will become less and less complete
and the effective contact hole area will be reduced by a
factor greater than k. The significance of the problems latent
here cannot be exaggerated.

## 3. Bonding defects

Let us leave the typical types of chip failure and turn to the
contact wires. Open bonds are the most frequent defect in this
area. They are somewhat under-represented in our distribution
of defect modes; at LSI chip areas with typically 2 contacts
per sq. mm of chip area, it is the chip defects that are the
center of attention. On the large number of SSI and MSI circuits with perhaps 10 contacts per sq. mm. of chip area, open
bonds have an overwhelming majority over other bond defects.
We shall only be discussing aluminum-pad gold-wire bonds as
no other type of contact enjoys a comparable degree of utilization.

Open bonds are often the result of manufacturing defects: contaminants or residual material on the aluminum pads, incorr-
ectly positioned bonds, double bonds etc. The inherent weak-
nesses of the nailhead bond are known since their invention
/16/17/. The intermetallic phases that ensure contact strength
are formed during bonding. This is not the end of the process,
however: even at 125°C, phase formation continues slowly.

High-quality Au
nailhead bonds have
exhibited lifetimes
of 30 years under
accelerating test
conditions, but
again and again
bonds are encounter-
ed whose lifetimes
are several orders
of magnitude
shorter, Figure 12.
It is well known
that elements such
as chlorine and
bromine shorten
lifetimes /18/.

(a)                    (b)

(a) good bond,   (b) degraded bond

**Au/Al Bonds**

Fig. 12

The role silicon plays in bond degradation is somewhat doubt-
ful. Quite a lot of elements and chemical compounds have been
suspected of being involved in degradation processes, but
in a few cases only such observations could be reproduced
under laboratory conditions. If the mechanism of degradation
had been really understood already we probably would have
been able to better allocate these observations.

It could well be that there is more than one mechanism
contributing. There is a evidence for several relatively
well- known metallurgical reactions /19/ plus one or more
corrosive processes and eventually a probably mechanically
acting inhibiting factor. In any case cavity forming processes are linked with bond degradation. Normaly they get
quickly explained by "Kirkendall voids", but undoubtedly
not all cavities are of this well understood origin. More
and more detailed investigations of this indispensable
bonding process are highly recommended.

## 4. Package related failures

Plastic packages are the most frequently used packages
for ICs. They suffer from an inherent weakness because the
plastic compound is permeable to water vapor. Particularly
low-power ICs, which do not heat above ambient temperature
or ICs, which are only intermittently used /20/, often show
corrosion of aluminum lines. This corrosion is caused by
the water, which diffuses through the plastic body, and by
corrosive ions, which may be positively or negatively
charged. Aluminum, because of its amphoteric character,
unfortunately is attacked by anions and cations. Generally
the ions, which cannot permeate the plastic body are assumed
to originate from contamination during manufacture.
This will certainly be true for the majority of cases, but
recently there has been discovered a new secret path for the
ions, which should be considered whenever corrosion took part
in a failure: As plastic compounds normally show bad adherence
to gold wires there is a circular capillary round the gold

wire. If in addit-
ion there has been
a gap opened between
plastic and lead-
frame ions may be
wandering into the
gap and reach the
chip via the capill-
ary tube along the
gold wire /21/.
Gaps are not so
unusual as might
be expected, but
they are not easily
detected. Figure 13
shows the traces
of liquids which
entered such gaps.
And in Figure 14
the plastic replica
of a ball bond is
to be seen with
the rest of the
gold wire extending
from it. A micro-
probe analysis
shows the distri-
bution of potassium,
which had as a
potassium cyanide
solution been in-
vading the chip
via the surface
of the gold wire.
That potassium cannot
be seen inside the hole simply has electron-optical reasons.

Corrosion

Fig. 13

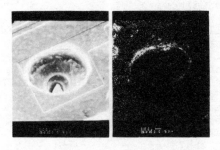

Plastic replica of nailhead        Potassium contamination   K$_\alpha$

Corrosion

Fig. 14

Package related failures will probably increase in number.
Large chip areas will create some headaches, as thermal
expansion coefficients of silicon, lead frames, passivating
layers, glue and solder layers cannot exactly be matched
and the risk of cracking the chip or part of it is dramati-
cally increasing /22/. New materials, sophisticated layers
with continually changing coefficients of expansion, elastic
or plastic wrappings of the chips will probably help. The
general problem will accompany IC engineer's work for some
while because chip areas will still be increasing. Chips
with areas of a centimeter square are already in production.


5. ESD Failures
Failures caused by electrostatic discharge are known as a
characteristic threat to MOS circuits. As integrated devices
become smaller and smaller, less and less energy is
necessary to destroy their elements.

Todays bipolar ICs are almost as susceptible to this mechanism as MOS circuits and the same countermeasures have to
be taken during manufacture and handling of such devices.
Protective on-chip devices, similar to those in MOS circuits,
urgently have to be developed for bipolar circuits.

We are used to
seeing all ESD
danger to ICs origi-
nating from charged
persons (human
body model) and
indeed this is a
main source of
charge and very
high voltages
(Figure 15). But
at least in hand-
ling MOS circuits
we are used to
carefully grounding
every person in-
volved. But there
is an additional danger coming from the ICs itself: With
growing pin numbers on plastic packages the capacitance

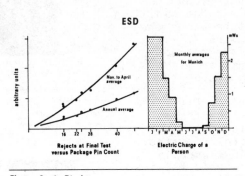

**Electro Static Discharge**

Fig. 15

of the package as
a whole is increased
and the charge it
can bear can easily
destroy small ele-
ments of the IC if
the leads of the
package are sudden-
ly grounded (charged
device model). If
this condition is
intentionally set
up for an experi-
ment, the depend-
ance of percent
failures on the
number of pins of

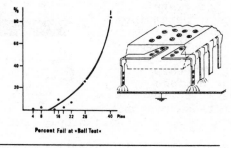

**Electro Static Discharge**

Fig. 16

the package as in Figure 16 is observed. In practice this
dangerous condition is frequently found when ICs are tribo-
electrically charged during their transport in production
or  after delivery,    and then get  into contact with well
grounded PC boards, vessels, etc. It is easy to understand
that this failure mechanism will increase in significance
when integrated elements continue to get smaller and
packages continue to exhibit more and more capacitance
/23/.

# 6.References

/1/ K.Yamabe, K.Taniguchi, Y.Matsushita, "Thickness Dependence of Dielectric, Breakdown Failure of Thermal $SiO_2$ Films", Proc. IRPS (1983) p. 180

/2/ J.R.Monkowski, R.T.Zahour, "Failure Mechanism in MOS Gates Resulting From Particulate Contamination", Proc. IRPS (1982) p. 244

/3/ H.R.Boling, "Process Defects and Effects on MOSFET Gate Reliability", Proc. IRPS (1980), p. 252

/4/ R.Coelho, "Physics of Dielectrics for the Engineer", Elsevier Scientific, (1979)

/5/ J.Fellinger, D.Koch, A.Schlemm, "Untersuchung des Durchbruchverhaltens von Oxiden in hochintegrierten MOS-Schaltungen", NTG Fachberichte 82 (1982) p. 97

/6/ E.S.Anolick, L.-Y.Chen, "Screening of Time-Dependent Dielectric Breakdowns", Proc. IRPS (1982),p. 238

/7/ W.K.Meyer, D.L.Crook, "Model for Oxide Wearout Due to Charge Trapping", Proc. IRPS (1983), p. 242

/8/ M.Itsumi, "Positive and Negative Charging of Thermally grown $SiO_2$ induced by Fowler-Nordheim Emission", J.Appl.Physics 52 (1981), p. 3491

/9/ A.Ito, H.A.Swasey, E.W.George, "Hot Electron Reliability Modeling in VLSI Devices", Proc. IRPS (1983), p.96

/10/ P.B.Ghate, "Electromigration-Induced Failures in VLSI Interconnects", Proc. IRPS (1982) p. 292

/11/ D.Pramanik, A.N.Sarena, "VLSI Metallization Using Aluminium and its Alloys, Part II", Solid State Techn., March 1983, p. 131

/12/ R.W.Pasco, J.A.Schwarz, "The Application of a Dynamic Technique to the Study of Electromigration Kinetics", Proc. IRPS (1983) p. 10

/13/ F.Fischer, F.Neppl, H.Oppolzer, U.Schwabe, "Electromigration Characterization of DC Magnetron-Sputtered Al-Si Metallization", Proc. IRPS (1983) p. 40

/14/ R.W.Thomas, D.W.Calabrese, "Phenomenological Observations on Electromigration", Proc. IRPS (1983) p. 1

/15/ P.A.Gargini, C.Tseng, M.H.Woods "Elimination of Silicon Electromigration in Contacts by the Use of an Interposed Barrier Metal", Proc. IRPS (1982) p. 66

/16/ E.Philofsky, "Intermetallic Formation in Gold-Aluminium
     Systems" Solid-State Electronics (1970) p. 1391

/17/ E.Philofsky, "Design Limits when Using Gold-Aluminium
     Bonds" 9th Annual Proc. Reliability Physics (1971)
     p. 114

/18/ R.C.Blish, L.Parobek, "Wire Bond Integrity Test Chip",
     Proc. IRPS (1983) p. 142

/19/ M.Kashiwabara, S.Hattori, "Formation of Al-Au Inter-
     metallic Compounds and Resistance Increase for Ultra-
     sonic Al Wire Bonding", Rev. Electr. Communic. Lab.,
     Vol. 17, No. 9 (1969) p. 1001

/20/ D.Ströhle, "Feuchteprobleme bei LSIs", NTG-Fachberichte
     82 (1982) p. 106

/21/ P.R.Engel, T.Corbett, W.Baerg, "A New Failure Mechanism
     of Bond Pad Corrosion in Plastic Encapsulated IC's
     under Temperature, Humidity and Bias Stress", 33rd
     ECC, (May 1983) Orlando, Fla.

/22/ H.Ullrich, "Probleme der Montagetechnik bei der Groß-
     integration", NTG-Fachberichte 80 (1980)

/23/ A.Walther, Siemens Internal Report (1983)

*Inst. Phys. Conf. Ser. No. 69*
*Paper presented at ESSDERC/SSSDT 1983, Canterbury 13–16 Sept. 1983*                                    121

# Liquid crystal optical processing

B. H. Soffer

Hughes Research Laboratories
3011 Malibu Canyon Road
Malibu, California 90265

Abstract.   Real-time nonlinear parallel processing techniques have
been implemented with liquid crystal light valves.   The transfer
function of the twisted nematic spatial light modulator can be
modified with half-tone masks in a coherent system to produce
arbitrary nonlinear functions.  Logarithmic transformation, allowing
separation of multiplicative images and subsequent homomorphic
filtering, was demonstrated. An A/D converter based on a pure
birefringent light valve using incoherent light was developed which
promises high throughput.  A device that maps local light intensities
to phase grating period variations was developed, which performs
nonlinear analog functions and digital optical logic.

Optical data processing has not fully achieved its potential of increased
capacity and speed compared with conventional electronic techniques, in
part for lack of a practical real-time spatial light modulator, and because
optical techniques have been almost exclusively limited to linear
operations.

I will review the research that we, along with our colleagues at the
University of Southern California, have done to implement real-time
non-linear parallel processing techniques employing liquid crystal light
valves developed and specially modified for these tasks.

Although it is well known that optical processing can provide speed and
throughput, it is interesting to compare various signal processing
techniques.  Some of these are shown on a time-bandwidth plot in Figure 1.
Liquid crystal light valves can be operated at time bandwidths of $10^6$.

The standard liquid crystal device is shown schematically in Figure 2.  A
biphenyl nematic liquid crystal mixture is operated in the field effect
mode.  The twisted nematic effect is combined with the optical
birefringence effect into the so-called "hybrid field effect" device.

The operating principle of the reflective mode device is shown in Figure 3.
The device is essentially a voltage divider that maps local light intensity
variations on a photoconductor into local voltage changes across the liquid
crystal.  The liquid crystal in turn modifies the state of polarization of

the readout light which is then analyzed into an intensity pattern replicating the original input. If the readout light is coherent then a coherent replica of the input image or data array is available for coherent optical processing. Fourier transforms, correlations and convolutions are readily done optically with coherent light.

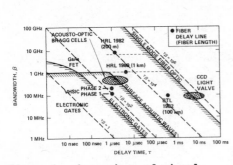

Figure 1.  Comparison of signal processing devices.

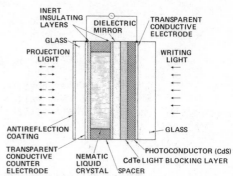

Figure 2.  Schematic of liquid crystal light valve

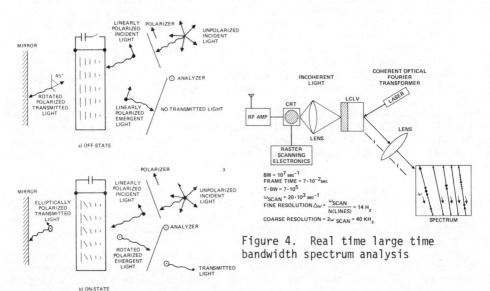

Figure 3.  Operating principle of hybrid field effect liquid crystal device

Figure 4.  Real time large time bandwidth spectrum analysis

An example of the use of coherent replication is shown in Figure 4.  A broad bandwidth RF signal of 10 MHz is raster scanned on an intensity modulated CRT.  A coherent replica of the raster is optically Fourier transformed to a high resolution raster in frequency with time bandwidth of $7 \times 10^5$.  The theoretical resolution of 14Hz was somewhat degraded by vibration and optical aberrations.  Replacing the liquid crystal light valve with its CCD electrically addressed version, under development, would eliminate the need for the CRT and input imaging lens.

A method of generating nonlinear functions, previously employed by Dashiell & Sawchuk at USC using photographic film, is the half tone screen technique. We replaced the film with the hybrid field effect device described earlier to create a real time implementation. The half tone screen converts a continuous tone image into a sampled version. Using a hard clipping or limiting recording medium such as lithographic film or a light valve, the samples are all equally intense but the spatial width of the samples depends on the spatial transmission of the half tone screen and the local intensity of the image. It is the analogy of pulse width modulation in communication. The nonlinear mapping of intensity to pulse width is mediated by the nature of the spatial transmission of the half tone screen.

To perform nonlinear function transformations a hard clipping, sharp threshold device with a linear, almost vertical transfer characteristic would be ideal. This ideal can be approached to some degree with a thicker CdS photoconductor layer in the LCLV. We have fabricated such a device and

have indicated the input/output characteristics of this thick CdS LCLV in Figure 5; in particular, the "on" and "off" states. The 200 Hz curve, which is somewhat steeper than the 2 kHz curve, shows a greater gamma. Most important is the remarkable difference in overall shape of the transfer characteristic. It is known that deviations from linearity and sharp cutoffs at the shoulders and toes will degrade device performance.

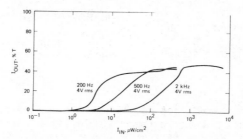

Figure 5. Transfer function of hybrid field effect LCLV

A custom-designed screen was used in conjunction with the LCLV to implement the logarithmic function to transform multiplicative images into additive ones. These images were then linearly optically filtered.

A half-tone screen designed for matching the transfer characteristics of a well-behaved light valve which showed little hysteresis was successfully used to generate the logarithmic function over a dynamic range of two orders of magnitude with an accuracy of +10%, or 1-1/2 decades with an accuracy of +5%. The original transfer function and the generated logarithmic transfer function are in the semi-logarithmic plots in Figure 6. Experiments were performed to demonstrate a real-time application of the logarithmic transformation by converting a multiplicative image into an additive one. The optical Fourier transform spectrum of two perpendicularly crossed (multiplied), coarse intensity gratings, tilted at 45° to the horizontal, is shown in the upper portion of

Figure 7. This was obtained by imaging the product of the grating transmission on the light valve input in its normal mode of operation. This is the expected diffraction pattern of two crossed gratings; "cross-terms" are visible in the corners of the square in the upper figure. The pattern is not simply the sum of the diffraction patterns of the two gratings. Operating the LCLV with a logarithmic transfer function (by inserting the half-tone mask in series with the LCLV) transforms the multiplicative spectrum into an additive one. The lower portion of Figure 7 shows the result of simply inserting the half-tone screen mask in series with the input imaged on the LCLV. The photographic exposure is equal in both the upper and lower parts of Figure 7. The 0th, 1st, and 2nd orders indicated are diffraction orders of the fine periodicity half-tone mask. In the 0th order, for which the mask was designed, cross terms are absent from the crossed grating diffraction, which demonstrates the transformation from multiplication to addition.

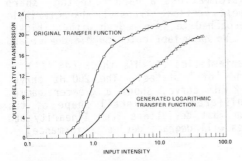

Figure 6. Liquid crystal light valve characteristic curve
(a) without the halftone screen
(b) with screen

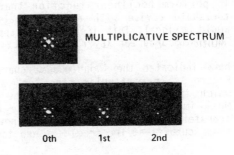

MULTIPLICATIVE SPECTRUM

0th     1st     2nd

LOGARITHMIC (ADDITIVE) SPECTRUM

Figure 7. Fourier spectra of crossed gratings

An experiment was performed that successfully demonstrated the concept of homomorphic filtering. A bar pattern simulating a video raster scanner noise was multiplied with a continuous tone image. This multiplicative noise can be seen as the vertical bars in Figure 8(a). After taking the logarithm, the noise is absent, the Fourier spectrum is greatly simplified, and the cross terms are removed. Now the bar pattern can be easily filtered out by the standard techniques of linear optical filtering in Fourier space. This process is called homomorphic filtering. The resulting filtered image in our example is shown in Figure 8(b).

As an example of performing nonlinear real-time optical processing directly with the nonlinearities of LCLVs, we have demonstrated a parallel A/D converter. No halftone screen techniques are needed in this application. The technique depends essentially on the quasi-sinusoidal transfer function of a "pure birefringent" or "tunable birefringence" LCLV. This A/D converter does not need coherent light, and no Fourier processing is done. The number of bits is limited by the aperiodicity of the transfer function.

The tunable birefringent light valve operation is employed in the parallel A/D converter. The intrinsic nonlinearity of the device is used to advantage for data processing.

(a) ORIGINAL OBJECT WITH MULTIPLICATIVE      (b) IMAGE AFTER HOMOMORPHIC FILTERING
    SCANNER NOISE

Figure 8.   Real-time homomorphic filtering

The large anisotropy of refractive index in nematic liquid crystals can be
utilized directly to make a field-effect device.  The diagram in Figure 9
illustrates the principle of the tunable-birefringence LCLV in which a
nematic of negative dielectric anisotropy is aligned with its director
essentially perpendicular to the electrode surfaces.  In the off-state,
plane-polarized incident light travels down the optic axis of the liquid
crystal, is reflected back, and is absorbed by a crossed polarizer
(analyzer).  When a field above a critical threshold voltage (typically
3 to 10 V) is applied, the liquid crystal begins to turn away from the
direction of the applied field, and then it no longer appears isotropic to
the normal incident light.  The liquid crystal film shows a birefringence
effect that increases with the applied voltage.  This introduces a large
phase retardation so that a given wavelength of light passes through the
analyzer in a series of maxima and minima as the voltage is increased.

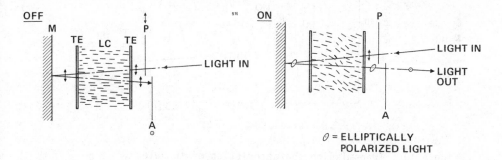

Figure 9.   Principle of tunable-birefringence LCLV

Generally, the transmission of a birefringent ($\Delta n$) layer between cross
polarizers is expressed as

$$T = \sin^2 2\phi \sin^2 \frac{\pi d \Delta n \sin^2 \theta}{\lambda},$$

where $\theta$ is the angle between the liquid-crystal optical axis and the incident light, $\phi$ is the angle between the input polarization and the direction of the tilt of the liquid-crystal optical axis, and d is the cell thickness. This optical effect that the liquid crystal has on the light beam has been treated as a function of the orientation of the liquid crystal molecules. However, if the liquid crystal has a positive dielectric anisotropy (the dielectric dipole moment is parallel to the optical axis), a voltage above a certain value will tilt the liquid-crystal molecules. This corresponds to a change in the angle $\theta$ (in the above case, from $90^\circ$ to lower values), which means a change in the optical retardation and thus a change in the transmission. In general, the optical transmission is a multiple peaked function of the applied voltage. This is shown in Figure 10 for the transmission of three different wavelengths in a test cell containing a Schiff base nematic mixture of N-(p-methoxybenzylidene)-p-butylaniline. The different colors, although not separated in their first transmission peak above threshold, become better separated at higher voltages. As a result, with incident white light, the viewer sees no light below $V_{th}$ at first. Then gray, white, and a sequential series of Newton colors appear as the cell voltage increases. The optimum transmission effects are obtained when the liquid-crystal director is realigned uniformly by the field with its $\phi$ angle at $45^\circ$ with respect to the incident polarized light. This can be accomplished by introducing a slight pre-tilt of the liquid crystal in that direction in the off-state. The more monochromatic the light, the better defined the modulation of transmission will be.

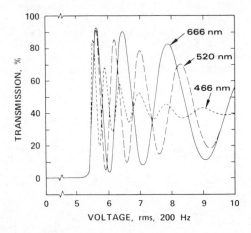

Figure 10.  Transmission characteristics of tunable birefringence cell between crossed polarizers. (Schiff-base mixture, perpendicular alignment)

The process of analog-to-digital (A/D) conversion is the representation of continuous analog information in sampled and quantized form.  For one-dimensional signals, any coding procedure that assigns to each signal value a group of digits (bits) performs such an operation.  For digital processing of multidimensional signals and images, the sampling, quantization, and digitization must be performed on an array of data.  We describe here a technique of real-time incoherent nonlinear optical processing that performs two-dimensional A/D conversion of images or other page-organized data in parallel without the need for scanning.  The method relies on the nonlinear characteristics of real-time optical input devices, such as the Hughes liquid-crystal light valve.  The overall relationship between the intensity transmittance of the device and the incident intensity at any point is given ideally by the sinusoidal curve shown with dashes in Figure 11(a).

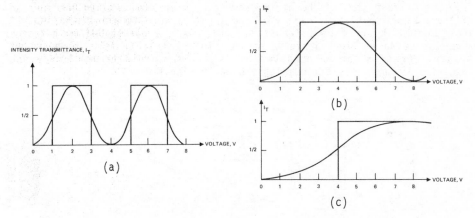

Figure 11.   Nonlinear characteristic curves for a three-bit grey code.  The step-function curves are desired for bit-plane outputs. The sinuous curves idealize the reponses of a birefringent device. Figures (a) through (c) represent increasingly significant bit-planes

The digital results of A/D conversion at each image point may be output serially as a bit sequence or in parallel as bit planes.  Bit planes are binary image planes, each of which displays the information of a particular significant bit of the digitized image.  The solid-line curves of Figure 11 show the nonlinear transfer characteristic needed to produce the bit planes of the three-bit reflected binary or Gray code and their relationship to the dashed curves of sinusoidal device characteristics.  When the output of Figure 11(a) is thresholded at one half, a 1 output is produced above threshold and a 0 output below, as shown by the curves with solid lines. This thresholding can be done electronically following light detection by a parallel array of sensors.  The threshold output in Figure 11(a) is the least significant bit of the three-bit Gray code.  The other two bits are obtained by attenuating the input intensity effectively to rescale the horizontal axis.  Use of the full dynamic range (0 to 8) gives the least significant bit.  Attenuating the input by a factor of 1/2 (to the range 0 to 4) gives the first cycle of the characteristic curve shown in Figure 11(b).  The last (most significant) bit is obtained by using an attenuation of one fourth so the curves of Figure 11(c) result.  Note that any continuous input between 0 and 8 gives a unique quantized three-bit output.  Although the thresholded outputs in Figure 11 are the three bits of the Gray code, other A/D code conversions, such as the usual straight binary code, can be achieved by translating these curves left or right along the horizontal axis. This can be done by introducing phase-retardation plates with different delays along orthogonal axes into the crossed polarizer system.

The system can produce these bits in parallel by placing an array of three periodically repeated attenuating strips over the write surface of the liquid crystal device, as shown schematically in Figure 12. The strips have attenuation factors of 1, 1/2, and 1/4, and the image of the strips is in register with a parallel photodetector array with electronic thresholding in the output plane. All three bits are sensed in parallel in this way. The period of the strips should be much smaller than the inverse of the maximum spatial frequency of the input picture to avoid aliasing. A two-dimensional array of attenuating spots with a corresponding detector array can also be used instead of the linear strip array. Simpler but slower operation can be achieved by using only one detector array and sequentially uniformly attenuating the entire input array.

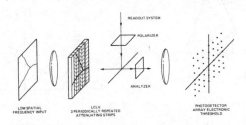

Figure 12.    System for parallel
A/D conversion

Although the theory behind the A/D conversion assumes a strictly periodic response characteristic, it is possible to produce the desired bit planes by using the quasi-periodic response curves of the actual device. The trade-off is that nonuniform quantization results.

There are no attenuating strips on the liquid-crystal device used in this experiment. Instead, the bit planes were generated serially. Also, the output was recorded on hard-clipping film rather than a thresholding detector array. A test target was generated that consisted of an eight-gray-level step tablet. The gray levels were chosen to match the quantization levels. This test object was imaged onto the liquid crystal device, and the output was photographically hard-clipped to produce the least significant bit plane of a three-bit A/D conversion. Next the write illumination intensity was decreased, effectively rescaling the response curve of the device to generate the next bit plane. The last bit plane (most significant bit) was obtained by attenuating the write intensity again and photographing the output. The input and the three-bit planes generated are shown in Figure 13.

Figure 13. Direct A/D conversion.
Top: eight level analog input.
Bottom: binary grey coded output
in 3 bit planes

Although the output contains some noise, the experiment illustrates the principle of real-time parallel incoherent optical A/D conversion. It was found later that the computer-generated gray scale was somewhat noisy because of the grain of the high-contrast film used. It is possible that future experiments with cleaner inputs and improved periodic light valves could produce better experimental results and more bits of quantization.

The potential A/D conversion rate can be estimated from typical parameters of currently available devices. The important parameters are device resolution (typically 40 cycles/mm), device size (typically 50 mm x 50 mm), and speed (generally 30 frames per second). Multiplying all these parameters together and dividing by 3 for the attenuating strips implies an A/D conversion rate of $4 \times 10^7$ points per second. A fully parallel system with one light valve and detector array for each bit plane could achieve $1.2 \times 10^8$ points per second.

The feasibility of real-time parallel A/D conversion on two-dimensional inputs has been shown in this experiment. In contrast with other techniques of real-time nonlinear optical processing, such as halftoning, the system operates with incoherent input. The requirements on the spatial and temporal coherence of the readout illumination are sufficiently relaxed that noise problems associated with coherent spatial filtering or transforming techniques are avoided. This technique also minimizes the spatial-frequency requirements of the real-time device because the sharp edges of the binary dots in halftoning do not have to be maintained.

We are developing a new class of liquid crystal light valves based on the variable grating mode (VGM) effect. In the VGM effect, certain LC materials exhibit a transparent phase grating whose local period is proportional to the locally applied voltage. When incorporated into a light valve sandwich structure, along with a photoconductive layer, local variations in the spatial distribution of the light signal or image intensity are converted into local variations of the grating period. In the Fourier transform plane, a novel additional parameter - the intensity - is coded into the spatial frequency distribution in a way that allows greater freedom to perform nonlinear optical data and image processing and to perform optical computations as well by simple spatial filtering and re-synthesis. We have examined many aspects of the device, including the LC system itself, the photo-conductive and electrical properties of the light valve, a new LC alignment scheme, and the detailed optical polarization properties of the VGM diffraction patterns. Furthermore, we have demonstrated several applications of the VGM device as well, both in analog nonlinear processing and in optical digital computation.

In the variable grating mode of operation of liquid crystals, a phase grating is formed with a period that depends upon the voltage placed across the cell. This phase grating originates from a variation of the optical path length due to a periodic orientational perturbation of the liquid crystal uniaxial index ellipsoid. The direction of periodicity is perpendicular to the quiescent state alignment of the liquid crystal molecules, i.e., the domains are parallel to the liquid crystal alignment in the off state. Typical spatial frequency variation is from 100 to 600 cycles/mm. Figure 14 shows typical voltage-induced behavior of a VGM cell as seen through a polarizing microscope. The period of the phase grating can be seen to decrease as the applied voltage increases.

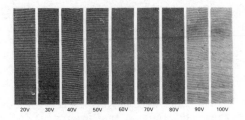

Figure 14.  VGM viewed through
polarizing microscope.

We have observed VGM domains in planar cells up to about 13 $\mu$m in thickness of the liquid crystal. These VGM domains for static fields are always parallel to the quiescent state alignment direction on the electrode surface and re observed only with applied dc fields. Our undoped liquid crystals were relatively high in resistivity ($>10^{10}$ ohm-cm) and showed little or no dynamic scattering even at dc voltages as high as five times the threshold voltage.

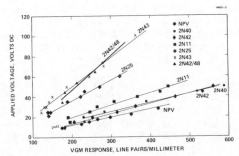

Figure 15.  VGM voltage dependence
for various liquid crystals.

The VGM effect in nematic-phase liquid crystals of negative dielectric anisotropy has been previously studied primarily with azoxy mixtures, such as Merck NP-V. These yellow-colored eutectic mixtures absorb light strongly in the near ultraviolet and blue (below 430 nm) region of the spectrum, and can undergo photodecomposition during extended illumination. We studied phenyl benzoate liquid crystal mixtures because they are colorless (strong absorption below 350 nm), are more stable to visible light exposures, and are more easily purified than the azoxy mixtures. Since we have ben studying structural effects (particularly molecular length) on the anisotropic and dynamic scattering properties of a series of phenyl benzoate mixtures, we examined the VGM response of this series as well as several other mixtures. The voltage dependence of their VGM domain grating frequency is shown in Figure 15.

Typical cells used in the foregoing experiments were fabricated from indium tin oxide (ITO)-coated, 1.27-cm or 0.32-cm-thick optical flats. The liquid crystal layer was confined by a 6 μm Mylar perimeter spacer. The liquid crystal surface alignment was obtained by spin-coating an aqueous solution containing polyvinyl alcohol on the ITO-coated surfaces, drying at $100^0$C, and gently rubbing to give a uniform directionality. The effect of dc voltage was observed with a polarizing microscope at 258X magnification for each liquid crystal mixture. The width d of the domain period (line pair) is inversely proportional to applied voltage according to

$$d = \alpha/V$$

where $\alpha$ is a constant that is dependent on the particular liquid crystal mixture. This relationship is illustrated in Figure 15, in which dc voltage as a function of grating frequency 1/d for each eutectic mixture is a straight line with slope $\alpha$. A smaller value of $\alpha$ is preferred, since a smaller slope gives a larger range of spatial frequency for the VGM effect per unit applied voltage. Thus, HRL-2N40 was chosen as the best of these ester liquid crystal mixtures for use in our studies of the photoactivated VGM device. The HRL-2N42 mixture is also of interest because its $\alpha$ value is almost as low as that of HRL-2N40 and its viscosity is considerably lower.

The grating lines in the VGM lie parallel to the surface alignment of the liquid crystal detector in the off state. Although rubbing has proven to be satisfactory for test cells, a much more uniform homogeneous alignment can be obtained by ion-beam etching certain types of surfaces. Figure 16 shows the best quality of domains that we have thus far obtained.

With the application of alternating fields of about 10 Hz, some cells were conductive enough to exhibit Williams-type domains. These domains are perpendicular to the quiescent state alignment and their periodicity is only slightly affected by the applied field. The Williams domains exist in a narrow range above the threshold voltage due to dynamic scattering resulting from increasing turbulent flow within the cell as the voltage is increased. This effect, in conjunction with the very small variation in the spacing of domains, limits the utility of this mode as a diffraction device. For frequencies below 10 Hz, a mixed-mode behavior is observed in the more conductive cells, with VGM and Williams-type domains appearing sequentially. The cells also exhibit severe scattering during the appearance of the Williams domains, especially as the frequency is raised. We have not been able to produce pure VGM behavior with alternating fields of zero average value even for cases of very asymmetric waveforms.

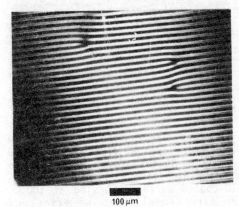

100 μm

Figure 16.  The phase grating structure of the VGM device at a fixed
voltage viewed through a phase
contrast microscope.

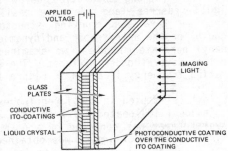

Figure 17.   Schematic diagram of the
VGM device construction.  Current
devices are read out in transmission
at a wavelength at which the photoconductor is insensitive.

The diffraction efficiency of the VGM cells depends strongly on the applied voltage, and can be as large as 25% in the second order (utilizing HRL2N40 in a 6 μm thick cell).   Interesting polarization effects of the odd and even diffraction orders as a function of the input light polarization are being investigated.  The structure of the present photoactivated VGM device is shown schematically in Figure 17.   The cell includes a vapor-deposited ZnS photoconductor and liquid crystal layer placed between ITO transparent electrodes that have been deposited on glass substrates.  In operation, the applied dc voltage is impressed across the electrodes.

The operating principle of the device is straightforward.   The photoconductor is designed to accept most of the drive voltage when not illuminated; the portion of the voltage that drops across the liquid crystal layer is below the activation threshold of the liquid crystal VGM effect.   Illumination incident upon a given area of the photoconductive layer reduces its resistance, thereby increasing the voltage drop across the liquid crystal layer and driving the liquid crystal into its activated state.   Thus, because of the VGM effect, the photoconductor converts an input intensity distribution into a local variation of the phase-grating spatial frequency.  The high lateral resistance of the thin photoconductive film prevents significant spreading of the photoconductivity and the associated liquid crystal electrooptic effect.  As a result, the light-activation process exhibits high resolution, as will be discussed in more detail below.

Because the VGM phenomenon is a dc instability effect occurring in high-resistivity ($\rho > 10^{10}$ Ω-cm) pure liquid crystal compounds, the device requires a high resistance photoconductive layer.   Zinc sulfide has been selected as the photoconductor material for the best resistance match with the liquid crystal layer.  Because the liquid crystal molecules are sensitive to photodecomposition in the ultraviolet, the ZnS layer is preferably made thick enough to optimize photoactivation in the blue region of the spectrum.

The ZnS layer is deposited on transparent ITO electrodes by evaporation or ion-beam sputtering methods. The sputtered films were 0.5 µm thick, highly transparent smooth surface layers. With the evaporation technique, we produced photoconductors of 1.5 to 5 µm thickness, characterized by a hazy, rough surface appearance that caused difficulties in liquid crystal alignment parallel to the electrodes. It has been reported that vaporized ZnS causes homeotropic or tilted homeotropic orientation of the liquid crystal material. Mechanical polishing of the evaporated photoconductors increased their transparency and surface uniformity, while polymer (PVA) coating the top of these ZnS layers, supplemented by additional surface treatment, resulted in good parallel alignment. Photoconductors were evaluated and compared by measuring the dark current and switching ratios of the resulting VGM devices.

From the preliminary photosensitive devices fabricated using a ZnS photoconductive layer to achieve the necessary high resistance, one cell was selected that aligned well and did not suffer the usual rapid deterioration seen in dc operation. This deterioration is assumed to result from poisoning of the liquid crystal by the diffusion of ions from the photoconductor. This particular cell was constructed of a 5 µm-thick evaporated ZnS layer that had been polished and then rubbed with surfactant polyvinylalcohol. The counterelectrode was an ITO transparent layer treated with the same surfactant. The 6 µm thick liquid crystal layer was made of HRL-2N40 ester. The dark series resistance of the 2.5 cm square cell was measured to be $3 \times 10^8 \, \Omega$. With -160V applied to the photoconductor electrode and with saturation illumination of 7.3 mW/cm$^2$ in the passband 410 to 550 nm, the spatial frequency of the VGM domains was calculated from the observed angles of diffracted orders to be 588 lines/mm. The device threshold at this illumination was 21 V, corresponding to a grating frequency of 103 lines/mm. The optical threshold at 160 V is of the order of 50 µW/cm$^2$.

Planar VGM test cells were studied with respect to edge effects on resolution and possible "spillover" of domains into adjacent unactivated areas. Electrodes were specially prepared by removing sections of the conductive coating by etching. A parallel plate cell was constructed such that there were conductive areas facing each other, either with conductive edges aligned or with a maximum overlap of 150 µm of a conductive electrode over the nonconducting area. Cell spacing was 6.3 µm and the material was merck NP-V. For alignment parallel to the edge when operating close to the threshold voltage, domains were parallel to the edge and within the active area. For higher voltages, there was fringe spillover by not more than one fringe spacing. For alignment perpendicular to the edge, domains appear to either terminate at the edge or to join with an adjacent domain.

The VGM liquid crystal device can be considered to be an intensity-to-spatial frequency converter capable of operating on two-dimensional images. The intensity-to-spatial frequency conversion allows the implementation of arbitrary point nonlinearities with simple Fourier plane filters. When an input image illuminates the photoconductor surface of this device the intensity variations of the input image change the local grating frequency. If coherent light is utilized to Fourier transform the processed image, different spatial frequency components of the encoded image, corresponding to different input intensities, appear at different locations in the Fourier plane as shown in Figure 18. Within the dynamic range of the device, intensities can thus be mapped monotonically into positions along a line in Fourier space. The input intensity distribution has thus been coded into Fourier (spatial frequency) space. If the spatial frequencies of the VGM domains are much larger than the largest spatial frequency component encountered in the images to be processed, we are in the tractable situation, familiar in communications theory, where the carrier frequency is much higher than the modulation frequencies. Thus, by placing appropriate spatial filters in the Fourier plane it is possible to obtain different transformations of the input

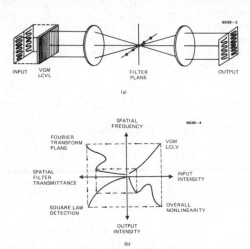

Figure 18. VGM nonlinear processing. (a) Experimental setup indicating the mapping of intensity to spatial frequency. (b) The overall input-output characteristic can be found by stepping through the successive nonlinear transformations including (1) the intensity to spatial frequency conversion, (2) spatial filtering, and (3) intensity detection.

intensity in the output plane as depicted in Figure 18(b). This figure describes the variable grating mode nonlinear processing algorithm graphically. The input intensity variation is converted to a spatial frequency variation by the characteristic function of the VGM device (upper right-hand quadrant). These variations are Fourier transformed by the optical system and the spectrum is modified by a filter in the Fourier plane (upper left-hand quadrant). Finally, a square-law detection produces the intensity observed in the output plane (lower left-hand quadrant). Considered together, these transformations yield the overall nonlinearity (lower right-hand quadrant). Design of a proper spatial filter for a desired transformation is a relatively easy task. For example, a level slice transformation requires only a simple slit that passes a certain frequency band or bands. A mathematical formulation of nonlinear processing using the VGM device shows

$$bv_0 \geq 2N$$

where $b$ is the pixel width, $v_0$ is the lowest usable VGM spatial frequency, and $N$ is the number of distinguishable grey levels. This relation requires that the pixel size contains $2N$ periods of the lowest grating frequency if $N$ grey levels are to be processed. For example, a 100 x 100 pixel image could be processed with 50 distinguishable grey levels on a 50 mm square device with $v_0$ = 200 cycles/mm.

The ability to perform arbitrary point nonlinearities over two-dimensional images greatly increases the flexibility of optical processing systems. In the past few years several different approaches to the problem of implementing generalized nonlinearities have been investigated. The main advantage of the VGM approach over previous methods is the ease of programming the functional nonlinearity desired for a given image transformation. This is done simply by changing the transmittance distribution of the spatial A filter in the optical processing system. The spatial filter is of relatively low resolution and need only have a space-bandwidth product equal to the number of gray levels to be processed independent of the space-bandwidth product of the input image.

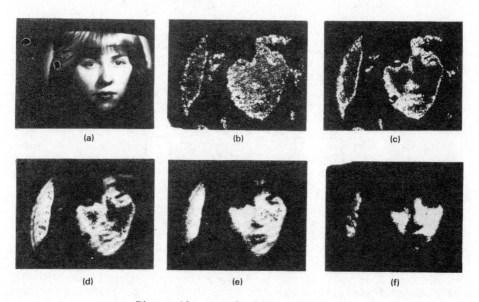

Figure 19. Level slice results

The same programmability advantage applies to the implementation of binary logic operations. One device can be used to implement any of the combinatorial logic operations (AND, OR, XOR, and their complements) by simply changing a Fourier plane filter. Previously described optical logic systems were "hardwired" to perform specific operations, and in most cases one or more logic functions proved difficult or cumbersome to implement. Use of the VGM liquid crystal device for the implementation of combinatorial logic operations is described below.

The ability of the VGM device to generate a level-slice nonlinearity was demonstrated. A continuous tone input picture is illuminated by an arc lamp source and imaged onto the photoconductor surface of a VGM device which initially exhibits a uniform phase grating structure due to a dc bias voltage. The grating period is locally modulated by the input picture intensity, and this modulation is mapped into a position along a line in the spatial filter plane. A red filter ensures that only the readout laser beam enters the coherent optical processor. Sectors of small circular annuli of varying radii are used to pass certain spatial frequency bands. This in effect allows only prescribed input intensity ranges to appear in the output. Circular rather than straight slits are used to capture the weak light which in small part is diffracted into circular arcs because of the grating imperfections. Figure 19 shows both the input and level sliced output pictures.

To see how the VGM device can be used to implement binary logic operations, one need only realize that the function of a logic circuit can be represented as a simple binary nonlinearity operating on the incoherent superposition of two binary images as input. As shown in Figure 20 NOT is simply a hard-clipping inverter, while AND and OR are hardclippers with different thresholds and XOR is a level slice function.

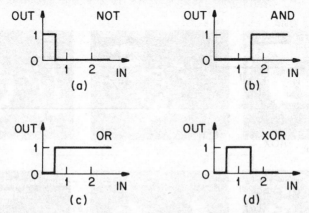

Figure 20. Logic functions as simple nonlinearities. Given an input consisting of the sum of two binary inputs, different logical operations can be effected on those inputs by means of the depicted nonlinear characterics. (a) NOT, (b) AND, (c) OR, (d) XOR.

The VGM device is well suited to implementing this type of nonlinearity. Since the nonlinearities associated with logic operations are binary functions, they can be implemented with simple slit apertures, i.e., 0 or 1 transmittance values. A noteworthy advantage of the VGM approach over previous optical logic methods is the ease of programming the nonlinearity, by merely changing the aperture in the spatial frequency plane.

Another feature of the VGM technique that is especially suitable for logic processing is that the input and output are physically separate beams. The input beam modulates a photoconductor; concurrently the image is read out with a second beam. This separation of input and output provides for the possibility of restoring the output levels to the desired 0 and 1 values even if the input levels are not exactly correct. This feature is essential to the production of a reliable logic system that is immune to noise and systematic errors in the levels. Electronic logic elements possess such level restoring capability, but currently proposed optical logic schemes lack this essential characteristic.

A series of experiments was conducted to demonstrate the fundamental logic functions. Two input fields were superimposed at the VGM plane along with a bias illumination. The total illumination intensity on the photoconductor of the VGM device was thus the sum of the two input

intensities and the bias intensity. The input illumination was filtered
high-pressure mercury arc lamp. The bias illumination was provided by a
collimated tungsten bulb source. The VGM device was read out in
transmission using a HeNe laser. A filter was placed in the Fourier plane
to select the desired spatial frequencies for each logic function.

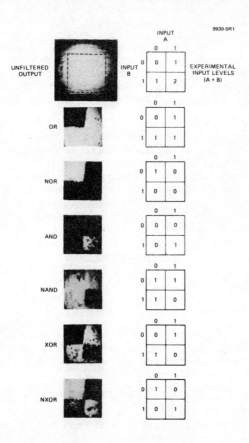

Figure 21. VGM logic results. The right-hand column indicates
ideal output levels for an image consisting of four quadrants
corresponding to truth table values. The left-hand column shows
the corresponding experimental results. The first row shows the
input for all experiments which consisted of a superposition of
two binary images. Succeeding rows show results for the logic
operations OR, NOR, AND, NAND, XOR, and NXOR, respectively.

For these experiments, the inputs consisted of one vertical rectangular
aperture and one horizontal aperture. When these were superimposed along
with the bias, a square image was formed with the four quadrants having the
intensity levels shown in Figure 21. This image corresponds to the logic
truth table shown. Thus the output images have intensity levels determined
by the truth table associated with the desired logic function. The logic
functions AND, OR, XOR and their complements were implemented sequentially

as shown in Figure 21 by altering only the Fourier plane filter. Imperfections visible to the output plane data arise from defects in the cell structure of the VGM device employed in these experiments.

The variable grating mode effect can be incorporated in a photoconductively addressed device structure which provides an overall intensity-to-spatial frequency conversion. The optical processing experiments using the VGM liquid crystal device described in this paper demonstrate the potential of this real-time optical image transducer for numerous parallel nonlinear operations on images. The molecular origin of the VGM phenomenon is now being studied in connection with the behaviour of the VGM phase grating as observed by polarization microscopy and polarization-dependent optical diffraction. Improvements in such characteristics of the device as response time, uniformity, dynamic range and density of defects are under continuing investigation.

This review represents the cooperative efforts of many of my associates at Hughes, including J. D. Margerum, A. M. Lackner, and D. Boswell, as well as my colleagues at the University of Southern California, A. R. Tanquay, T. C. Strand, A. A. Sawchuk, and P. Chavel. It is based on work sponsored by the Air Force Office of Scientific Research under Grant AFOSR-77-3285 at USC and Contract F49620-77-C-0080 at HRL. Liquid crystal work on the project was partially supported by the Air Force Office of Scientific Research on Contract F49620-77-C-0017 at HRL.

BIBLIOGRAPHY

Variable Grating Mode Effects

  L. K. Vistin, Sov. Phys. Dokl. 15, 908 (1971)

  W. Greubel and W. Wolf. Appl. Phys. Lett. 19, 213 (1971)

  J. M. Pollack and J. B. Flannery, in Liquid Crystals and Ordered Fluids J. J. Johnson and R. E. Porter, eds. (Plenum Press, New York 1978) 2, 557.

  J. M. Pollack and J. B. Flannery, Society for Information Display 1976 Intern. Symp. Digest. 143 (1976)

  P. K. Watson, J. M. Pollack and J. B. Flannery in Liquid Crystals and Ordered Fluids, J. F. Johnson and R. E. Porter, eds. (Plenum Press, New York, 1978), 3, 421.

  M. I. Barnik, L. M. Blinov, A. N. Trufanov and B. A. Umanski, J. de Physique, 39, 26 (1978).

  B. H. Soffer, D. Boswell, A. M. Lackner, P. Chavel, A. A. Sawchuk, T. C. Strand and A. R. Tanguay, Jr., Proc. Soc. Photo-Opt. Instrum. Eng. 232, 128 (1980).

  P. Chavel, A. A. Sawchuk, T. C. Strand, A. R. Tanguay, Jr. and B. H. Soffer, Opt. Lett. 5, 398 (1980).

Nonlinear Optical Processing

  H. Kato and J. W. Goodman, Appl. Opt., 14, 1813 (1975)

  T. C. Strand, Opt. Commun, 15, 60 (1975).

  S. R. Dashiell and A. A. Sawchuk, Appl. Opt., 16, 1009 (1977).

S. R. Dashiell and A. A. Sawchuk, Appl. Opt. 16, 2279 and 2394 (1977)

B. J. Bartholomew and S. H. Lee, Appl. Opt. 19, 201 (1980).

A. Tai, T. Chang and F. T. S. Yu, Appl. Opt. 16, 2559 (1977)

Real-time Nonlinear Optical Processing.

D. Casasent, Opt. Eng. 19, 228 (1974).

S. Iwasa and J. Feinleib, Opt. Eng. 13, 235 (1974).

A. Armand, A. A. Sawchuk, T. C. Strand, D. Boswell and B. H. Soffer, Opt. Lett., 5, 129 (1980).

J. D. Michaelson and A. A. Sawchuk, Proc. Soc. Photo Opt. Instrum. Engin., 218, 107 (1980).

Optical Logic.

R. A. Athale and S. H. Lee. Opt. Eng., 18, 513 (1979).

S. A. Collins, Jr., U. H. Gerlach and Z. M. Zakman, Proc. Soc. Phot. Opt. Instrum. Eng., 185, 36 (1979).

D. H. Schaefer and J. P. Strong, III. Proc. IEEE, 65, 129 (1977).

L. Goldberg and S. H. Lee, Appl. Opt., 18, 2045 (1979).

*Inst. Phys. Conf. Ser. No. 69*
*Paper presented at ESSDERC/SSSDT 1983, Canterbury 13–16 Sept. 1983*

# Interconnect and contact technologies for VLSI applications

P. A. Gargini

Intel Corporation, Santa Clara, CA, U.S.A.

Abstract. Device scaling has been implemented throughout the chipmaking industry as a means of increasing density and performance. However, progressively larger chips require longer interconnections that degrade signal propagation delay. In addition, it becomes increasingly harder to make low resistance contacts to shallow junctions without inducing shorts to the substrate. To solve these problems, technologists are advancing multiple layers of interconnection and contact barrier techniques in VLSI processes.

## 1. Introduction

Device scaling is being extensively used to simultaneously improve chip density and device performance. Advances in lithographic and plasma etching techniques are projected as reaching submicron line capability before the end of the decade (Fig. 1). In addition, improved manufacturing procedures are continually reducing the number of defects per $cm^2$, thus making larger dice sizes possible with the subsequent advantages of increasing the number of functions and decreasing cost (Fig. 2). An understanding of the physical laws regulating device scaling has enabled reliable transistors to be produced with progressively thinner gate oxides and shallower junctions. However, the key element of the Integrated Circuit era remains the capability of contacting and interconnecting all the

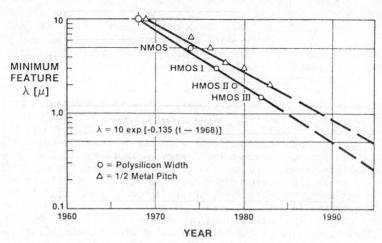

Fig. 1 Minimum device feature vs time

devices on the same chip through a few technological steps. If the wiring capability had not come with the semiconductor technology, the VLSI revolution would have probably stopped far short of the 1K memory chip. As predicted by historical trends, commercial devices with more than one million transistors will be in high volume manufacturing well before the year 1990 when chips containing more than 4

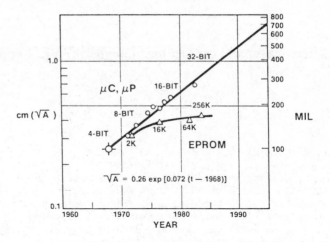

Fig. 2 Trend of chip size vs time

million transistors will be introduced. However, it is becoming increasingly clear that major changes in contacting and interconnecting techniques are essential for the achievement of these chip density levels. Multiple layers of low resistivity materials will become the standard way to interconnect densely packed devices along with diffusion barriers placed in the contacts to prevent silicon-aluminum interdiffusion either during fabrication or the expected lifetime of a part. It is the purpose of this paper to review limitations, problems and possible solutions in the development of interconnections and contacts for VLSI applications.

## 2. Effect of Scaling Interconnections on Circuit Performance

Interconnections are used in Integrated Circuits to carry electrical signals from one device to another. In principle, interconnections should not affect overall circuit performance as it should be controlled by transistor drive and switching capabilities as well as specific circuit design techniques. However, physical dimensions, material properties, and interaction with surrounding structures control how electrical signals propagate along interconnection lines. Practical interconnection lines introduce a finite time delay in the propagation of electrical signals across an integrated circuit (1,2). In addition, capacitive coupling between adjacent interconnections introduces unwanted signal noises that must be taken into account during the design stages.

The theory on scaling down device dimensions was introduced more than 10 years ago (3,4). The transistor scaling relations are shown in Table I for both the constant field and the constant voltage conditions. The former has so far represented an ideal situation, as supply voltage has remained fixed at 5 volts, whereas the latter is representative of a more realistic implementation. In any case, it is clear from Table I that gate delay ($\tau_g$) decreases as device dimensions are reduced. The effect of scaling on interconnect properties, as predicted by the original scaling theory, is reported in Table II under the restrictive assumption that the capacitance between the interconnection and the substrate can be modeled as a parallel plate capacitor. No interaction with adjacent lines is included. The key elements of this simplified model and the relations between the rise time of the voltage at the end of the line and the time constant of the interconnect line are shown in Fig. 3. Table II shows that the characteristic time constant of an interconnect line does not change as a result of the scaling operation for either the constant field or constant voltage case. However, reduced line cross section causes the normalized voltage drop across the line and the current density to increase. The first effect degrades noise margins whereas the second is a reliability concern.

## 3. Practical Scaling of Interconnections

It is important to understand that scaling has been used in two different ways in the I.C. industry. The first application consists in an optical shrinkage of an existing layout. In this case the primary goal is to increase the total number of good dice per wafer and is primarily a cost reduction operation. Improved device performance is a consequence. On the other hand, scaling is also used to increase device density in order to produce larger memory or logic circuits. Although the effect of scaling on transistor performance is identical in the two cases, the effect on interconnections is drastically different. In the former case, interconnect length decreases according to the scaling laws, whereas in the latter case, interconnect length remains constant or increases. Scaling parameters for the second case are shown in Table III. In both cases (i.e., for die shrinkage and larger chips) other deviations from the basic scaling theory have occurred. In order to minimize the voltage drop along interconnect lines, the conductor thickness has been kept essentially constant. Moreover, to partially offset the effect of increased interconnection length on signal propagation delay, the field oxide thickness has decreased at a slower rate than is predicted in theory. This situation is summarized in the last column of Tables II and III.

In conclusion, horizontal device scaling has been applied to all lateral dimensions. Vertical scaling has been applied to transistor gate oxide and junction depth but only slightly to the thickness of conductors and field oxide regions. As a result, signal propagation delay along practical interconnect lines decreases for die shrinkage and increases for increased number of devices on a larger chip.

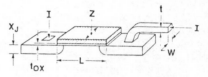

| | CONSTANT FIELD | CONSTANT VOLTAGE |
|---|---|---|
| L, Z, t, W, $x_J$, $t_{OX}$ | $\frac{1}{k}$ | $\frac{1}{k}$ |
| Voltage $V_S$, $V_{TH}$ | $\frac{1}{k}$ | 1 (constant) |
| Current I | $\frac{1}{k}$ | k |
| Power VI | $\frac{1}{k^2}$ | k |
| Gate Delay $\frac{CV}{I}$ | $\frac{1}{k}$ | $\frac{1}{k^2}$ |
| Power Delay $CV^2$ | $\frac{1}{k^3}$ | $\frac{1}{k}$ |

## SCALING FACTOR k > 1

Table I Scaling relationships of MOS transistors

| CIRCUIT PARAMETER | CONSTANT FIELD | CONSTANT VOLTAGE | |
|---|---|---|---|
| | | | t ~ CONST, h ~ CONST |
| $R_L \quad \rho \frac{L}{w\,t}$ | K | K | ~ 1 |
| $C_{LS} \quad t \frac{w\,L}{h}$ | 1/K | 1/K | ~ 1/K$^2$ |
| $T_L : R_L\,C_{LS}$ | 1 | 1 | ~ 1/K$^2$ |
| $\frac{V_L}{V} \quad \frac{I_L\,R_L}{V}$ | K | K$^2$ | ~ K |
| $J_L - \frac{I_L}{w\,t}$ | K | K$^3$ | ~ K$^2$ |

Table II Effect of scaling and reduced die size on interconnections

It is important to realize that the above considerations do not take into account any fringing field (line to ground) or line to line capacitance. Several publications have addressed these limitations through proposed capacitance models that more closely approximate real interconnections (5, 6, 7, 8). These effects are particularly important as line to line spacing becomes comparable with line width and line to substrate dimensions. A more precise capacitance model providing for the

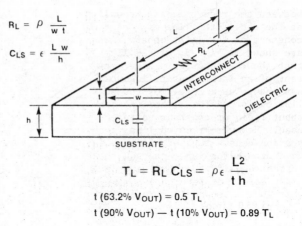

$$R_L = \rho \frac{L}{w\,t}$$

$$C_{LS} = \epsilon \frac{L\,w}{h}$$

$$T_L = R_L\,C_{LS} = \rho\epsilon \frac{L^2}{t\,h}$$

$$t\,(63.2\%\ V_{OUT}) = 0.5\ T_L$$

$$t\,(90\%\ V_{OUT}) - t\,(10\%\ V_{OUT}) = 0.89\ T_L$$

Fig. 3 Simplified interconnect model

above effects was developed to calculate line propagation delay. The capacitance model used gives results consistent with those reported in reference (7). A study of devices commercially available in recent years reveals that conductors and dielectrics have started to substantially scale down around the 3μ range of the minimum feature size λ. Empirical expressions for metal and polysilicon width, space and thickness as well as dielectric thicknesses as a function of λ are shown in Fig. 4. It is clear that the only unknowns in the bottom equation of Fig. 4 are λ and Lmax once the conductor resistivity (ρ) and the dielectric constant (ε) have been selected. By using the historical expressions of A (Fig. 2) and λ (Fig. 1) and the assumptiom of Lmax $\sim\sqrt{A}/2$ for metal interconnections and Lmax $\sim\sqrt{A}/4$ for polysilicon interconnections, the relation between $\tau_L$ and λ can be generated (Fig. 5). Experimental values of τg are reported for comparison. The higher 4 points are representative of typical NMOS/HMOS I, II, III values whereas the lower points obtained on transistors defined by X-ray lithography are obtained by references (9) and (10). It is assumed for simpli-

| CIRCUIT PARAMETER | CONSTANT FIELD | CONSTANT VOLTAGE | |
|---|---|---|---|
| | | I – CONST. h – CONST | |
| $R_L = \rho \frac{L_{MAX}}{w\,t}$ | $MK^2$ | $MK^2$ | $\sim MK$ |
| $C_{LS} = \epsilon \frac{L_{MAX}\,w}{h}$ | $M$ | $M$ | $\sim M/K$ |
| $T_L = R_L\,C_{LS}$ | $M^2K^2$ | $M^2K^2$ | $\sim M^2$ |
| $\frac{V_L}{V} = \frac{I_L\,R_L}{V}$ | $MK^2$ | $MK^3$ | $\sim MK^2$ |
| $J_L = \frac{I_L}{w\,t}$ | $K$ | $K^3$ | $\sim K^2$ |

Table III Effect of scaling and increased die size on interconnections
M is the scaling factor for long-range interconnections

city that a circuit is not interconnect limited if $\tau_L < \frac{1}{2}\tau g$. It can be seen that signal propagation delay of polysilicon lines has become a problem below ~3μ minimum feature whereas aluminum has not yet become a limiting factor. Clearly, in order to reduce the adverse effects of polysilicon resistivity on signal propagation delay, it is necessary to substitute or add a new material capable of satisfying all the processing requirements met by the polysilicon and yet presenting a lower resistivity. These attempts have been made since the late sixties and are the object of the following paragraphs.

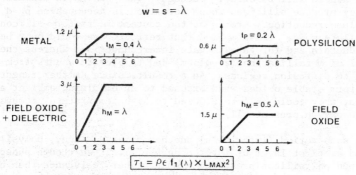

$$w = s = \lambda$$

$$T_L = \rho \epsilon\ f_1\ (\lambda) \times L_{MAX}{}^2$$

Fig. 4 Scaling assumptions

## 4. Gate Conductors (Mid-Range Interconnections)

The first MOS processes introduced into production in the late '60's, used
pure aluminum as a gate conductor. Since aluminum cannot withstand the
high temperatures of diffusion and oxidation processes, its use required
that source and drain regions were in place before aluminum deposition.
As a consequence, the source and drain overlaps of the gate region were
dictated by alignment requirements with a consequent detrimental effect on
transistor performance (high source and drain to gate capacitance). The
use of aluminum as interconnection and as gate electrode made wiring very
easy. The contact mask was allowing access of the aluminum to diffusion
regions. However, high resistance diffusions were the only conductors
available to cross with aluminum connectors. To overcome this problem,
several attempts to introduce an additional level of aluminum were made.
The typical low temperature dielectric available at the time was APCVD
(atmospheric pressure chemical vapor deposition) silicon dioxide which was
primarily used for scratch protection. Unfortunately, the step coverage
properties of this oxide were very marginal and any subsequent aluminum
deposition suffered from thinning (or opening) at steps whenever the second
aluminum was crossing the first aluminum level. It was then proposed that
a metal capable of with-
standing the high tem-
perature of a more con-
formal silicon nitride
deposition should replace
the first layer of alu-
minum. Refractory ma-
terials such as tungsten
and molybdenum, already in
use as backup plates for
silicon power devices,
were proposed due to their
high melting points.
Bulk resistivity of these
materials is about 5μΩ cm.
However, several problems
prevented this solution
from succeeding. First
of all, refractory ma-
terials do not reduce
silicon dioxide (alumi-
num does), consequently,

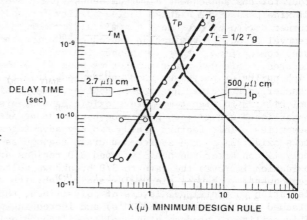

Fig. 5 Comparison of gate and
line delay vs minimum feature

they do not adhere well to oxide. Secondly, when heated in contact with silicon, they produce silicides whose formation is accompanied by a substantial volume reduction. Cracks at the contact refractory-silicon were a common failure mode. This implied that refractories could only be used as gate conductors and for gate to gate interconnections whereas the second level of metallization (aluminum) had to provide shunts from gate connections to diffusion regions. As a result, since neither tungsten or molybdenum form stable oxides when exposed to an oxidizing ambient at high temperatures, they could only be exposed to high temperature processes if there was no oxygen present (11).

During this same period (late '60's) another material (i.e., polysilicon) was proposed as first layer of interconnect (12). Even though phosphorus or boron doped polysilicon presented a much higher resistance than either molybdenum or tungsten (not to mention aluminum) it immediately showed several properties that were bound to make it the ruler of the seventies. Polysilicon can easily be deposited by CVD techniques (i.e., good step coverage), oxidizes without producing unstable oxides, adheres well to silicon dioxide, and, most importantly, it can stop diffusions or implants from substantially degrading the gate oxide while source and drain regions are produced in a self-aligned mode. Figure 5 clearly shows that in the seventies transistor performance rather than interconnect delay controlled the overall signal propagation delay across an I.C. chip. Several attempts were made to upgrade the pure molybdenum technology so as to compete with the silicon gate process. In principle, molybdenum should stop any phosphorus, boron or arsenic ion because of its high mass number. However, molybdenum tends to deposit in a columnar structure that allows ion implanted ions to easily pass through this and the gate oxide thus reaching the channel region. To eliminate this effect, additional C.V.D. oxide deposition was required and/or the photo-resist layer used to define the gate conductor was left in place to prevent source and drain ions from penetrating the channel region. The technique of phosphorus doping the C.V.D. oxide in order to round the steps during the subsequent high temperature processes was used for both polysilicon and pure refractories. Phosphorus doped silicon dioxide partially flows when exposed to high temperature (around $1000^{\circ}$C or so), thereby insuring an easy-to-cover contour for the subsequent aluminum deposition. Phosphorus doped silicon dioxide presents a high gettering ability for sodium impurities. This turned out to be extremely beneficial for molybdenum because it allowed removal of any contamination introduced during the molybdenum deposition step coming from either molybdenum source itself or from handling procedures. This was never a major problem for polysilicon because of the high purity of the gases used and the possibility to expose the deposition chamber to HCl cleaning steps. However, the possibility of exposing the polysilicon to high temperature oxidizing ambients was the key factor that established the silicon gate process as the uncontested dominator of the seventies. This feature yields two key advantages. First of all, it allows oxidation of the source and drain regions as well as the edges of the polysilicon after the heavily doped S/D regions are put in place. This oxidation improves the gate to S/D breakdown voltage substantially and consequently enhances die yields and reliability. Secondly, it provides isolation to additional layers of polysilicon, thus making possible novel stacked gate structures (EPROM's) and increasing packing density of DRAM's. This ability to produce high quality silicon dioxide when exposed to high temperature oxidizing steps more than offsets the lower resistivity of molybdenum in practical MOS devices. It was not until the late seventies that a massive effort to overcome the resistivity limits of the polysilicon was launched.

## 5. Silicides

Silicides of Pt and Pd have been used since the early seventies to produce
Schottky barriers and ohmic contacts in low power TTL gates made with bipolar devices. The use of a Schottky diode between the base and collector
regions of a bipolar device reduces the amount of charge stored in the
base region and in this way reduces the switching time of the transistor
going from the saturated to the off conditions. In addition, the silicides'
delay of any unwanted interdiffusion between silicon and aluminum (13)
allows ohmic contacts to be more reliably formed to shallow diffusion
regions. Typically, silicides present higher resistivities than the
corresponding transition metals (14). However, several silicides such as
$MoSi_2$, $WSi_2$, $TaSi_2$, $TiSi_2$, $NiSi_2$ and $CoSi_2$ form a stable layer of silicon
dioxide when exposed to a high temperature oxidizing ambient so long as
a suitable source of silicon atoms is available typically below the
silicide film (15). The thickness and growth kinetics of the silicon
dioxide film is very similar to the growth of oxide on polysilicon film.
This property was the missing link in the polysilicon upgrade path and
the reason for the shift in the research effort from pure refractories
to their silicides. As in the early days of polysilicon, several deposition techniques were (and still are) attempted to deposit stable silicides. In any case, the intrinsic difficulty of dealing with a compound
as opposed to a single material makes the silicides a full step more complex
than polysilicon. The first approach consisted of depositing the compound
by coevaporation techniques. This deposition procedure has the advantage
of using highly pure elements. Substantial technological problems had to
be overcome in this case due to the high melting point of the refractories
typically utilized and to the poor heat conduction properties of silicon.
This latter limitation required the development of a scanned e-beam technique to prevent "silicon splashing." A more suitable technique was cosputtering of the individual elements. In this case, any problem related
to the high melting point of transition metals or poor heat conduction of
silicon is overcome by the sputtering technique. Practically, cosputtering
is best used in the development stages but once the optimum "silicide
composition" is determined, sputtering from a single preformed target is in
general preferred for production purposes. In both cases, substrate
cleanliness and impurity inclusions during deposition are critical in determining physical properties of the silicide layer during subsequent processing steps. Several problems have so far prevented the use of silicide
as a pure replacement of polysilicon. First of all, adhesion of silicides
to oxide is not as strong as polysilicon to oxide. Secondly, step coverage
of either evaporation or sputtering technique is not as conformal as CVD
which is presently the only technique used to deposit polysilicon (Hot
wall LPCVD has become the standard for polysilicon since the second half
of the seventies). Thirdly, the interface silicide-silicon dioxide
properties are not as well understood as the polysilicon-silicon dioxide
interface. And finally, most silicides decompose in oxidizing ambient
unless a suitable source of silicon atoms replenishes the silicon consumed
to produce silicon dioxide. All of the above reasons have led to the use
of silicide in a sandwich structure with polysilicon.

## 6. Polycide

The term polycide, from the words polysilicon and silicide, indicates a
sandwich of silicide on top of polysilicon. This approach has several

advantages over the pure silicide approach and is emerging as the solution
of the '80's to reduce resistance of the gate conductor. In this approach,
the polysilicon film is deposited and doped as in the standard silicon gate
process (lower doping levels are acceptable in this case). In early attempts to produce a silicide layer onto the polysilicon, researchers
deposited refractory metals directly on top of the polysilicon layer and
then reacted the two materials to form the polycide sandwich. Silicon is
the predominant diffusing species in most cases and its diffusion is very
sensitive to small amounts of oxide present on the polysilicon surface.
Typically, non-planar silicide-polysilicon interfaces are observed after
sintering and full polysilicon grains might dissolve during silicide formation. Silicide formation is accompanied by a substantial reduction in
volume (20-30%) and generates high tensile stress levels. Even though
this technique yields the lowest silicide resistivity reported, the above
problems have led to the use of deposited silicides instead of silicides
obtained by metal with polysilicon. After the silicide film is deposited
on top of the polysilicon and annealed to reduce the resistivity, the
sandwich structure is defined. The subsequent steps follow the standard
silicon gate approach. This structure fully overcomes the problems outlined in the previous paragraph. First of all, adhesion between polysilicon and silicides is in fact superior to adhesion between silicides and
silicon dioxide. Secondly, polysilicon presents very good step coverage
and even if the silicide film is discontinuous at steps, the underlying
polysilicon insures gate conductor continuity. Thirdly, the well-known
properties of the polysilicon-silicon dioxide interface are essentially
preserved unchanged by the polycide structure with consequent threshold
stability and proven reliability. Finally, the underlayer of polysilicon
provides a source of silicon atoms for the silicide during oxidation steps,
and, if things are "done right", a silicon dioxide film will form on the
silicide without decomposing it. However, several precautions must still
be taken in handling silicide films even when used in a polycide structure.
Stresses in all refractory silicides are quite high (around $1-2 \times 10^{10}$
dynes/cm$^2$) and tensile in nature. The mismatch between the thermal expansion coefficient of silicon and silicide accounts for most of the stress
generated during the cool down from the sintering operation. Plasma etch
techniques have been successfully used to etch the most common silicides.
Fluorine mixtures have been used to etch silicides of molybdenum, tungsten
and tantalum whereas chlorine based plasmas have been used for titanium.
Additional care must be taken during wet precleans in the case of $TiSi_2$
because it reacts energetically with HF based solutions.

Recent investigations of low pressure chemical vapor deposition of silicides have had very encouraging results (16, 17, 18). This deposition
technique offers the advantage of providing both very pure films and
excellent step coverage. Very low resistivity has been reported for these
films after annealing due to virtual elimination of the impurities plaguing
other deposition techniques. It should be remembered that CVD of polysilicon was a major breakthrough in the silicon gate technology.

## 7. Salicide

As line dimensions scale down to micron and submicron dimensions, even
though polycide gate conductors alleviate signal propagation delay problems, the high sheet resistance (>100$\Omega$/$\square$) of diffusion regions becomes a
major problem. Diffusion regions are in fact useful for short range interconnections. In addition, as diffusions are scaled to .25$\mu$ or less, the
voltage drop between the contact to the source and drain regions and the

electrical channel substantially degrades transistor output characteristics. The Self-Aligned silicide structures (SALICIDE) are formed after following the standard silicon gate process flow up to polysilicon etching (19). A sidewall spacer to insure adequate source/drain to gate isolation and S/D to channel separation is then formed. The spacer also has the advantage of preventing the heavy N+ and P+ implants from reaching the channel region that is bridged to source and drain by a light implant placed before spacer formation. This arrangement reduces hot electron effects. Once the spacer is formed, a thin layer (200-500 Å) of transition metal is deposited across the wafer. Silicide formation is then induced in the silicon and polysilicon regions but not on top of the oxide regions. This excess metal is then removed by an appropriate etch that does not attack the silicide. Silicides in which the metal is the dominant moving species are formed by thermal annealing. With silicon as the dominant moving species, outdiffusion from both S/D and polysilicon regions creates silicide on the spacer region and consequently causes shorting to the transistor electrodes. By taking appropriate precautions, $TiSi_2$ has been successfully used in salicide applications even though silicon is the dominant diffusing species during silicide formation. Transition materials where silicon is the dominant diffusing species can still be used if Ion Beam Induced mixing is used to form the silicide. Since the IBI technique induces silicide formation at a low temperature, all problems associated with silicon diffusion are overcome. It should be pointed out that a substantial amount of silicon is consumed to produce the silicide. Typically, the silicide reaction consumes a thickness of 500 to 1100 Å of silicon to produce 1000Å of silicide. As junction depth is further scaled down, this silicon consumption will not be tolerated. However, some materials like Pd or Pt generate a snowplow effect in the underlying As implant that prevents shorting of the junction to the substrate (20, 21). Some structures completely replacing source and drain regions by Schottky diodes made by silicide on silicon have also been reported. The problem of choosing the silicide material can easily be solved for p channel devices. On the contrary, it has been very difficult to make good Schottky diodes with P regions. However, lower transistor gain than the usual S/D diffused transistors has been observed.

## 8. Alternative Low Resistance Gate Structures

Polycides are the most common approach pursued in reducing sheet resistance of gate interconnections because they can be easily retrofitted in a std MOS process. It is clear that as chip size increases and consequently interconnect lines become longer, the lower resistivity of pure metals will make them more attractive than silicides. In addition, process temperatures will be reduced to 800°C or less in order to limit lateral diffusion of source and drain regions. Conformal dielectric deposited by plasma techniques at temperatures below 500°C and dielectric planarization techniques are already available to replace the std dielectrics. These observations indicate that the oxidation properties and high melting points of silicides will not constitute a critical factor in the future. An attempt to use a pure metal in a low temperature process has been reported in which tungsten was selectively deposited on silicon (or polysilicon) to form interconnections (22, 23). In this approach, the transistor gate and interconnections are defined in a silicon nitride/polysilicon sandwich. The source and drain regions are formed and the sides of the polysilicon oxidized. The nitride layer is then removed from the top of the polysilicon and tungsten is selectively deposited. Low temperatures (< 500°C) are then used in the deposition of the dielectric (plasma enhanced CVD) and in the subsequent processing steps. The final gate structure closely resembles a polycide structure. Furthermore, a structure similar to the salicide approach can be obtained if a spacer is used to isolate the polysilicon side walls. This restricts the only high tempera- ture step after gate oxidation to source and drain activation. This is just an example suggesting that as technologists become more expert in low temperature processing techniques, transition metals and eventually alumi- num will make their way back into the gate structure to minimize signal propagation delay of mid-range interconnections (Fig. 6).

## 9. Long-Range Interconnect Materials

Any material used as long-range interconnection in I.C. should meet sever- al requirements:
1. Low resistivity
2. Adaptable to practical methods of deposition
3. Good adhesion to typical dielectric
4. Low resistance contacts to both p+ and n+ type silicon ( or to materials which satisfy this requirement)
5. Electromigration resistant
6. Patternable into fine lines
7. Compatible with bonding techniques.

Of the four materials which present the lowest resistivity (Table IV), only aluminum satisfies (within limits) all the above requirements (25).

Aluminum adheres very well to the dielectrics com- monly used in I. C. manufacturing whereas Au, Cu and Ag need an inter- mediate layer to act as a glue. Cr, Mo, Ti:W, Ti-Pt and Ti-Pd have been successfully used as buf- fer layers (24). From a manufacturing point of view, aluminum is very easy to work with since it can be depo- sited by either evap- oration (melting point $660^{\circ}$C) or sputtering techniques. In addition, wet chemical and plasma etching procedures for patterning fine alumin- um lines are well esta- blished. Aluminum and silicon do not form any compounds, and the phase diagram between two ele- ments is very simple. At the eutectic temperature ($577^{\circ}$C) only 1.6 atomic percent of silicon is dis- solved into aluminum.

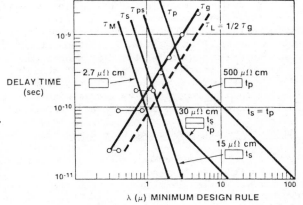

Fig. 6 Signal propagation delay of several interconnect materials vs minimum feature

| METAL | Al | Au | Cu | Ag |
|---|---|---|---|---|
| $\rho_{BULK}$ [$\mu\Omega$ - cm] | 2.70 | 2.38 | 1.69 | 1.62 |

Table IV Resistivity of long- range interconnect materials

Aluminum contact resistance is typically in the $10^{-6} - 10^{-7} \Omega\text{-cm}^2$ range for both p+ and n+ regions. Gold wire bonds are commonly made to aluminum thin film metallization in assembling semiconductor devices. Five intermetallic phases will form if Au and Al are heated above $250^{\circ}$C (26). Typically, $Au_5\text{-}Al_2$ is the dominant phase. Kirkendall voids may possibly nucleate if temperature and time during and after bonding are not carefully chosen. Thermal cycling could reduce the time to failure by as much as one order of magnitude.

## 10. Multiple Level Interconnections

Interconnections occupy a large fraction of the die area. For random logic typically as much as 40% of the die is occupied by interconnections. In order to increase the packing density, multiple layers of metallization have been introduced. The simplest scheme consists in the $Al/SiO_2/Al$ system. The choice of aluminum limits processing temperature to below $450^{\circ}$C. Double layer of aluminum reduces to about 25% the area occupied by interconnections for random logic. In addition, it allows reduction of the average interconnect length and thereby also reduces the average signal propagation delay. However, metal crossovers introduce coupling noise between the two layers of aluminum that must be taken into account during the design phase. The second layer of aluminum is thicker than the first since it has to cover a more complex topology without substantial thinning down on steps. Typically the first layer of aluminum is used to contact silicon regions whereas the second only contacts the first layer of aluminum. This is due to the fact that the thickness of the intermetal dielectric is comparable to the thickness of the dielectric below the first layer of aluminum. The implication is that substantial overetch of the via holes to the first metal occurs in order to reach the silicon regions. Overetch of via holes enlarges lateral dimensions and, if the via contact becomes larger than the first metal landing pad, step coverage problems affect the second metal. Aluminum is the almost universally used conductor for multiple layer applications. Addition of Cu, Ni, Mg and Mn to inhibit hillock formation in the first layer of aluminum and thus reduce dielectric defects while improving electromigration resistance has been reported. However, Au and W have also been used for both first and second layer of interconnections (27, 28). The electromigration resistance of both materials is better than Al. Silicon dioxide has been commonly used as an interlayer dielectric. Sputter-etching techniques have been used in order to partially planarize the $SiO_2$ surface (29). The simultaneous sputter-etching of the surface during quartz deposition creates a unique surface topology on top of first metal landing pads. This allows a layout of non-overlapping vias without risking step coverage problems for the subsequent aluminum deposition. In recent years, polyimides have been proposed as multilevel insulators (30). Polyimides are applied in liquid form like photoresists and then cured to remove the solvent. They provide an excellent planarization of the surface but a number of process related problems has limited their success.

## 11. Contacts

Completed integrated transistors require contact regions to the different electrodes to interconnect devices. The scaling theory predicts that resistance/contact will increase as $K^2$ as window area is reduced. This implies that contacts will degrade transistor performance due to a voltage drop across them that will reduce the voltage drive across the electrical channel. Aluminum has brilliantly solved the wiring problem of the I.C. industry for the last two decades because of its ability to function as an interconnect material and to make low resistance contacts to heavily doped p+ or n+ regions. It is well known that a thin (10 to 20 Å) layer of native oxide quickly forms on the surface of the silicon as soon as it is exposed to air. This oxide layer if left in place would yield high resistance contacts. However, aluminum reacts with silicon dioxide when the two materials are heated in intimate contact. Therefore, if time and temperature are appropriately chosen, aluminum will reduce the native silicon dioxide and will alloy with the underlying silicon. At temperatures below 500°C, typical of MOS or bipolar manufacturing after aluminum deposition, the solubility of silicon in  aluminum is small. In fact, if the silicon were uniformly removed from the contact region, then less than 100Å of silicon would be required to reach the solid solubility limit in 1μ of aluminum. However, deep pits in the silicon can be observed if the aluminum is removed after the sintering operation. This indicates that all the silicon is provided by specific sites and not by the full contact surface. Possibly, the uneven nature of the native oxide is responsible for enhanced silicon dissolution at preferential locations where the oxide is thinner. Once the silicon starts to diffuse into the aluminum the remaining vacancies are backfilled by aluminum. Silicon diffuses very fast into thin aluminum films. (D~$10^{-8}$ cm$^2$/sec @ 450°C). Therefore, once the silicon enters the aluminum film, it quickly moves away from the contact region, thereby causing more silicon to dissolve in order to satisfy the solid solubility at the sintering temperature. Since the aluminum keeps backfilling the silicon voids, an electrical short to the substrate eventually develops. The depth of the dissolution pits is strongly dependent on contact size (increasing as contact size decreases), and it imposes a lower limit to practically achievable junction depths (31). To work around the problem the MOS industry has chosen to provide deeper junctions beneath the contacts and shallower junctions in all the other regions. This solution is obviously not usable for bipolar transistors due to the verical implementation of the device. Addition of silicon (1-3%) to the aluminum has proven effective in further reducing silicon dissolution in contacts. In order to reduce the detrimental effects of fast silicon diffusion on contact integrity, a metallization scheme including a doped polysilicon/aluminum sandwich has been reported. The polysilicon provides a distributed source of silicon for the aluminum line which eliminates enhanced silicon dissolution in the contact region. This solution is quite simple to implement for NMOS processes but requires n+ and p+ polysilicon for CMOS applications, and it does reduce reliability of interconnect lines. A more desirable solution consists in interposing an additional layer(s) of material between the aluminum and the silicon to completely eliminate any interdiffusion phenomenon (32).

## 12. Contact Metallization Schemes

To produce chemically and electrically stable contacts, the contact material(s) must meet the following goals:
1.  produce low resistance contacts
2.  produce a chemically stable system with silicon
3.  produce a chemically stable system with aluminum.

Silicides of metals have been extensively investigated because several of them sufficiently meet the first two goals. Transition metals form silicides when in contact with silicon by solid-solid chemical reaction without forming a liquid phase. This is very important in obtaining a uniform and controlled interface. Contact resistance between silicides and silicon is typically in the $10^{-7}$ $\Omega cm^2$ range while contact resistance between aluminum and silicides is typically in the $10^{-8}$ range. Typically, silicide reactions begin between $200°C$ and $600°C$ depending upon whether metal rich silicides, monosilicide or disilicides are formed. The metal and/or (most commonly) the silicon diffuse into the other and start to react. Subsequent silicide formation proceeds by the dominant diffusing species moving through the silicide and reaching the other material until all the metal is completely consumed. The most silicon rich silicide is the most stable from a thermodynamic point of view. As long as the junction is deeper than the silicide-silicon interface, junction integrity is preserved and electrically good ohmic contacts are achieved. These results have been reported with many silicides. The silicide-silicon system is typically quite stable over a wide temperature range. However, the stability of the silicide-aluminum system must also be carefully understood because aluminum still remains the simplest interconnect choice. When aluminum is heated in contact with noble or near noble metal silicides at temperatures around $400°C$, it decomposes the silicides and triggers a chain of events which quickly leads to the failure of the contact (33). On the other hand, refractory metal silicides are more stable in contact with aluminum. $MoSi_2$ was found to be stable in contact with aluminum up to $500°C$. However, silicon and aluminum diffuse through the refractory silicides. Failure of $TiSi_2$ to act as a silicon/aluminum diffusion barrier has been documented. Pure transition metals have been used as diffusion barriers also. Ti, V and W react very little with aluminum for temperatures in the $400-450°C$ range and are fairly impervious to silicon diffusions as long as some unreacted metal remains (34, 35). Good ohmic contacts have been obtained with all of them. However, they do not form any compound or eutectic when heated in contact with silicon. Therefore, the interface properties are largely controlled by the cleanliness of the deposition step. In an attempt to combine the stable interface properties of silicides and the lower reaction rate of aluminum and transition metals, several multilayer metallization schemes have been proposed. Tungsten, Ti:W (10:90) and TiN appear to be the most effective barriers whereas silicides of platinum and palladium have been widely used to produce good electrical contacts. It should be pointed out that it is not straightforward how silicides and barriers should be combined on the basis of experiments in which testing of the silicide or barrier properties were independently done. For instance, titanium acts as an efficient barrier if placed between aluminum and silicon. 1000 Å of titanium will survive 30 min anneal at $450°C$ if the reaction is perfectly uniform because only 600Å will have been converted to $TiAl_3$. $TiAl_3$ is not a diffusion barrier for either Si or Al. Also, little interaction occurs between Ti and $Pd_2Si$ if heated below $500°C$ (36). However, if titanium is interposed between aluminum and $Pd_2Si$ drastic interactions are observed even after 30 min at $400°C$. This is due to diffusion of aluminum through the titanium which reacts with the silicide. In addition, the silicon from the silicide, Ti and Pd further react to form a ternary phase. Some Pd reaches the Al layer. These effects could not be predicted from the aforementioned experiments. In fact, when Al diffuses through Ti and reaches the Si interface no reaction occurs and the small solubility of Al in Si will limit the diffusion rate through the Ti. On the contrary, $Pd_2Si$ acts as a sink of aluminum due to the fast interaction between the two materials thus accelerating aluminum diffusion. The salicide structures described in a previous paragraph reduce aluminum-to-silicon effective contact resistance by approximately the ratio of the silicide-silicon area to the aluminum-silicide area. In fact, aluminum to silicide contact resistance is about an order of magnitude lower than that of silicide to silicon.

## 13. Electromigration

Electromigration is defined as the directional motion of ions of the conducting material resulting from a momentum exchange with current carrying
electrons. Aluminum lines under D.C. bias conditions fail due to their
becoming open circuits because of electromigration of aluminum in aluminum. A large body of publications is available on the subject. The Mean
Time to Failure for open lines has the following form:

$$MTTF = A(wt/J^2) \exp(E_a/KT)$$

Where w is the line width, t the line thickness, J the current density, A
a factor embodying several physical properties, K the Bolzmann constant and
T the absolute temperature. $E_a$ represents the activation energy for the
electromigration phenomenon. It has been experimentally proven that $E_a$
varies from about .5ev for small grained aluminum to about 1.2 ev for large
grained aluminum covered with a glass overcoat (37). The two activation
energies are indicative of grain boundary aluminum diffusion and bulk
aluminum diffusion respectively. The addition of copper to aluminum thin
film strips has brought about a considerable increase in lifetime without
substantially changing film resistivity. This MTTF increase is attributed
to absorption of copper atoms on aluminum grain boundaries with resultant
impediment to aluminum grain boundary diffusion (38). Since copper atoms
are smaller than aluminum atoms, more rapid diffusion of copper occurs.
Subsequent to copper depletion, migration of aluminum starts. It has been
shown that MTTF increases as copper content increases up to 4-8% by weight
(39). However, solubility of copper in aluminum is very small (less than
.1% @ $250^0$C) so that any improvement in MTTF should quickly saturate for
copper concentrations much less than 1%. Replenishment of copper in grain
boundaries from $CuAl_2$ precipitates has been proposed to explain the improvement of MTTF with copper concentration. MTTF decreases when copper concentration reaches values beyond 8%. This has been explained by observing
that $CuAl_2$ clusters are more probable as copper concentration increases.
It is reasonable that clusters would be less effective sources of copper
since the surface to volume ratio is less than with individual precipitates.
An increase in MTTF has also been observed when magnesium is added to
aluminum. The previous equation has been proven incorrect for lines of
small physical dimensions. As line width is decreased as compared to
median grain size, MTTF is increased. This effect is due to a reduction
of potential sites for atomic or vacancy flux divergence which in turn
reduces the probability of electromigration failure. It should be pointed
out, however, that minimum width lines are seldom subjected to the highest
current density. On the contrary, wide metal bus lines are bound to carry
more current as more devices are placed on a chip or more current is required as more powerful transistors are obtained from a die shrinkage
operation.

Both aluminum and silicon electromigration in contacts has been reported
(40). The silicon is removed from the contact region due to a momentum
exchange with the electron flux going from the silicon into the aluminum.
In this case, as in the case of aluminum spiking during contact formation
described in paragraph #11, aluminum backfills the voids left by the
silicon until a short to the substrate develops. The activation energy
for silicon electromigration is about .9-1.1ev. This value is in agreement
with the activation energy reported for diffusion of silicon in thin aluminum films.

Electromigration studies of electrically good contacts (3 X 3μ in size,
1.5μ junction depth under the contact) stressed under high current (10mA)

and high temperature ($250^{\circ}$C) conditions have shown that silicon electro-migration at the contacts and consequent aluminum shorting to the sub-strate is the only failure mode (Ea~.9ev). On the other hand, as stress temperature is reduced to $125^{\circ}$C (or less), all the failures are due to open circuits. These failures result from aluminum electromigration at the edge of the contact (Ea~.5ev). The MTTF for the two failure modes has been extensively studied by subjecting several contacts of different physical parameters to stress under a wide variety of current and temperature conditions (41). The conclusions of this study can be summarized as follows:

1. contact always fails before metal lines
2. silicon electromigration (leakage failures) dominates under high temperature, high current stress conditions
3. aluminum electromigration (open failures) dominates under low current, low temperature conditions
4. silicon voids are located at the leading edge of the contact
5. current flows into the contacts mostly from the periphery
6. MTTF for leakage failures is proportional to $X_J^2$.

Integration of more transistors on a die will increase chip temperature and, as specified by device scaling theory, current density into the con-tacts will increase as $K^3$ (constant voltage). This implies that leakage failures will eventually become dominant in the normal operating range. Interposition of diffusion barriers such as tungsten or TiN (42) have proven completely effective in suppressing silicon failures.

It was shown in the previous paragraph that interaction between aluminum and silicides or transition metals is largely responsible for the instab-ility of the contact-interconnect system once silicon migration has been minimized through the use of the above materials. All the interactions between thin films are primarily occurring through diffusion of materials along grain boundaries which can be several orders of magnitude faster than diffusion in bulk materials. As explained in this paragraph, alu-minum electromigration resistance can be substantially increased by addition of copper which is absorbed on aluminum grain boundaries thus reducing/suppressing aluminum migration along these paths. It could be inferred that as the interconnect material (Al) is optimized with respect to electromigration resistance by reducing (through Cu) grain boundary diffusion of aluminum, the aluminum-barrier interaction should also benefit since interdiffusion of the two materials must play a role in the reaction rate of the intermetallic formed. This has been proven true for both Ti and W in contact with Al-Cu (43). In both cases a substantial reduction in intermetallic formation has been observed even though the detailed reaction between Al-Cu and Ti or W was quite different in nature. These findings further indicate that minimal changes in the interconnect-contact materials drastically affect the stability of the system.

## 14. Topology

As device features are further reduced in the 1-2$\mu$ range and less, new topological problems must be solved as lateral and vertical dimensions become comparable. These effects are particularly pronounced in the rear end of the process where contacts, vias and interconnections are made and accumulation of multiple films maximizes vertical dimensions while interfering geometries create unique surface contours.

Aluminum step coverage of contacts has been traditionally optimized by partially flowing (at temperatures ~800°C) the C.V.D. dielectric (doped with an appropriate impurity) in order to round all the sharp corners. Similar problems encountered with step coverage of vias have been solved in the past by creating a slope in the via sidewall through appropriate etch techniques as dielectric reflow cannot be used if the first layer of interconnection is composed of aluminum.  In both cases a trade off between aluminum step coverage and minimum allowable spacing of contact to adjacent structure must be made as indicated in figure 7.  These results are consistent with those previously published in reference (44). A new contact/via metallization procedure has been proposed which would solve both maximum step coverage and minimum dimension of contact/via requirements.  The approach consists in creating metal studs, before dielectric deposition in the regions where contact should be made, by using lift off techniques.  Once the dielectric is deposited the surface is spun with thick photoresist which produces a flat topology.  Both photoresist and subsequently the dielectric are etched back with a plasma etch step which has a selectivity of 1:1 between the two materials.  Once the resist is completely removed, leaving a flat dielectric surface where the top part of the metal studs is barely emerging, the etch process is stopped.  Aluminum deposition then occurs on a flat topology (45).

## 15.  Conclusions

The scientific research conducted in the 60's led to the fortunate discovery of the silicon/silicon dioxide stable interface properties as well as the diffusion masking ability of silicon dioxide with respect to boron, phosphorus  and arsenic.  The high temperature oxidation properties of polysilicon and the ability of aluminum to serve as both an interconnect material and a low resistance contact to $n^+$ or $p^+$ silicon made them the natural choice to complete the selection of materials presently used in the silicon gate process.  The use of silicon nitride as self-aligned oxidation mask (LOCOS) and the flow properties of phosphorus  doped oxide solved the topological problems of the technology. The seventies were characterized by the engineering exploitation of the above combination of compatible materials which led to the conception and realization of novel devices.  Increased chip density was achieved by decreasing device features in accordance with the scaling theory. However, although transistor performance progressively improved as a result of device miniaturization, and interconnect performance also improved for die shrinkage applications, the latter remained constant or degraded as number of functions/chip were increased.  The only way to mimimize the adverse effects of interconnect scaling on circuit performance consisted in decreasing both interconnect and field oxide thicknesses at a slower rate than predicted by the scaling theory.  This approach aggravated topological problems and imposed more stringent requirements on aluminum and polysilicon etching operations.

As minimum device feature approached the 3μ range, signal propagation delay of polysilicon lines became comparable to gate delay thus significantly contributing to total signal propagation time.  A massive effort to overcome polysilicon limitations was launched in the late seventies spurring a renewal of basic scientific activity in the silicon technology. Transition metals and even more successfully their silicides proved to be an effective way of reducing resistivity of mid-range interconnections. Polycide and salicide structures are emerging as the natural upgrades of polysilicon interconnections. As low temperature process requirements and technological advancements proceed, pure transition metals and eventually aluminum will make their way back into a composite gate structure to maximize signal speed.

Contact resistance and contact integrity were also negatively affected by the scaling operation. Salicide structures are effective in reducing overall aluminum to silicon contact resistance. However, interaction of aluminum with the silicide or diffusion of silicon through it pose severe reliability concerns. Interposition of diffusion barriers between aluminum and silicide constitute the engineering solution to the overall contact integrity problem.

Aluminum still remains the most effective long-range interconnect as adhesion problems affect Au, Cu and Ag which have lower resistivity than aluminum. These require buffer layers to stick to commonly used dielectrics. Eventually, as more materials are introduced into silicon technology to overcome limitations imposed by scaling on the standard constituents of the silicon gate process, the addition of buffer layers will not appear a substantial penalty in order to reduce interconnect resistance. Of course, the introduction of the above materials will require more effort and understanding than the oversimplified picture described here. In the immediate future, addition of additives such as copper to aluminum will substantially improve electromigration resistance at the expense of a minimal increase in film resistivity. In addition, absorption of copper on aluminum grain boundaries may further reduce interaction between aluminum and transition metals (or their compounds) used as diffusion barriers.

Polycides and salicides as well as multiple layers of aluminum interposed with planarized dielectrics will emerge in the 80's to produce devices with as many as 4 million transistors and minimal signal propagation delay.

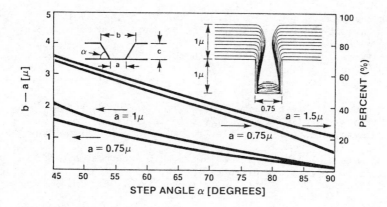

Fig. 7 Aluminum step coverage of small size contacts

References

1.  K C Saraswat and F Mohammadi, IEEE Trans. on EL. Devices, vol. ED-29, no. 4, p 645, Apr. 1982.
2.  A K Sinha, J A Cooper, Jr., and H J Levinstein, IEEE EL. Dev. Lett., vol. EDL-3, no. 4, p 90, Apr. 1982.
3.  R H Dennard, F H Gaensslen, L Kuhn, H N Yu, Tech. Digest IEDM, Washington, D. C., p 168, Dec. 1972.
4.  R H Dennard, F H Gaensslen, H Yu, V L Rideout, E Bassous, and A R LeBlank, IEEE J. Solid-State Circuits, vol. SC-9, p 256, 1974.
5.  A E Ruehli and P A Brennan, IEEE J. Solic-State Circuits, vol. SC-10, no. 6, p 530, Dec. 1975.
6.  A E Ruehli, IBM J. Res. Devlop., vol. 23, no. 6, p 626, Nov. 1979.
7.  R L M Dang and N Shigyo, IEEE Electron Device Lett., vol. EDL-2, p 196-197, Aug. 1981.
8.  W. H. Chang, IEEE Trans. Microwave Theory Tech., vol. MTT-24, p 608-611, Sept. 1976.
9.  P I Suciv, E N Fuls, and H J Boll, IEEE EL. Dev. Lett., EDL-1, p 10, 1980.
10. A K Sinha, VLSI Science and Technology/1982, vol. 82-7, p 173, 1982.
11. D M Brown, W E Engler, M Garfinkel, and P V Gray, J. Electrochem. Soc., 115, no. 8, p 874, 1968.
12. F Faggin and T Klein, Solid State Electronics, vol. 13, p 1125, 1970.
13. C. J. Kircher, J. of Appl. Phys., vol. 47, no. 12, p 5394, Dec. 1976.
14. S P Murar Ka, M H Read, C J Doherty, and D B Fraser, J. Electrochem. Soc., 129, no. 2, p 293, Feb. 1982.
15. M Bartur and M-A Nicolet, Appl. Phys. Lett., 40(2), p 175, Jan. 1982.
16. D L Brors, J A Fair, K A Morning, and K C Saraswat, Solid State Technology, p 183, Apr. 1983.
17. W I Lehrer, J M Pierce, E Goo, and S Justi, Fall Meeting, Electrochem. Soc., abs. 165, 1982.
18. S Inoue, N Toyokura, T Nakamura, M Maeda, and M Takagi, Jap. Semic. Tech. News, vol. 1, no. 3, p 25, 1982.
19. C M Osburn, M Y Tsai, S Roberts, C J Lucchese, and C Y Ting, VLSI Science and Technology/1982, vol. 82-7, p 213, 1982.
20. M Wittmer, C-Y Ting, and K N Tu, J. Appl. Phys., 54(2), p 699, Feb. 1983.
21. R L Thornton, El. Lett., vol. 17, no. 14, p 485, 1981.
22. P A Gargini and I Bienglass, 169th Meeting, Electrochem. Soc., Denver, abs. 381, 1981
23. P A Gargini and I Beinglass, Tech. Dig. IEDM, Washington, D.C., p 54, Dec. 1981.
24. P B Ghate, J C Blair, and C R Fuller, Thin Solid Films, 45, p 69, 1977.
25. A J Learn, J. Electrochem. Soc., vol. 123, no. 6, p 894, 1976.
26. E Philofsky, Solid St. El., 13, p 1391, 1970.
27. D Summers, Spring Meeting Electrochem. Soc., San Francisco, abs. 435, 1983.
28. J Mikkelson, L A Hall, A K Malhotra, S D Seccombe, and M S Wilson, IEEE J. Solid State Circuits, SC 16, p 542, 1981.
29, J S Lechaton, Fall Meeting Electrochem. Soc., Los Angeles, abs. 585, 1979.
30. L B Rothman, J. Electrochem. Soc., vol. 127, p 2216, 1980.
31. S Vaidya, J. of Electronic Materials, vol. 10, no. 2, p337- 1981.
32. M-A Nicolet, Thin Solid Films, 52, p 415, 1978.
33. G J van Gurp, J L C Daans, A van Ostrom, L J M Augustus, and Y. Tamminga, J. Appl. Phys., 50(11), p 6915, Nov. 1979.

References:

34.  K Nakamura, S S Lau, M-A Nicolet, and J W Mayer, Appl. Phys. Letters, vol. 28, no. 5, p 277, 1976.
35.  M Bartur and M-A Nicolet, Thin Solid Films, 91, p 89, 1982.
36.  T G Finstad and M-A Nicolet, Thin Solid Films, 68, p 393 (1980).
37.  J R Black, Proc. of the IEEE, vol. 57, no. 9, p 1587, 1969.
38.  I Arne, F M d'Heurle, R E Horstmann, IBM J. Res. Develop., 14, p 461, 1970.
39.  A J Learn, J. of Electronic Materials, vol. 3, no. 2, p 531, 1974.
40.  J R Black, I.R.P.S., San Diego, p 233, 1978.
41.  P A Gargini, C Tseng, and M H Woods, I.R.P.S., San Diego, p 66, 1982.
42.  C Y Ying, J. VAC. Sci. Technol., 21 p 14, 1982.
43.  I Krafcsik, C J Palmstrom, J Gyulai, E Colgan, E C Zingu, and J W Mayer, Spring Meeting Electrochem. Soc., San Francisco, abs. 436, 1983.
44.  I A Blech and H A Vander Plas, J. Appl. Phys., 54(6), p 3489, 1983.
45.  L B Rothman, J. Electrochem. Soc., vol. 130, no. 5, p 1131, 1983.

*Inst. Phys. Conf. Ser. No. 69*
*Paper presented at ESSDERC/SSSDT 1983, Canterbury 13–16 Sept. 1983*

# Integrated circuit process modelling

C Hill and A L Butler

Plessey Research (Caswell) Ltd., Allen Clark Research Centre,
Caswell, Towcester, Northants., NN12 8EQ

Abstract   The use of existing computer programmes to model integrated circuit fabrication processes is reviewed, and the particular features of future programmes capable of modelling VLSI fabrication steps are described.  Emphasis is placed on the need for good basic experimental data in ion-implantation, diffusion and oxidation of small geometry one- and two-dimensional structures in amorphous, single and polycrystalline silicon.  Some new or improved experimental techniques suitable to obtain this data are described.

## 1. Introduction

The goal of process modelling is to relate quantitatively the final device structure to the relevant variables of the fabrication process. Necessarily, such modelling has been an essential part of integrated circuit technology development from the beginning; but only in the past few years have comprehensive process models embodied in computer programmes appeared, enabling rapid and quantitative simulation of the cumulative effects of the large number of sequential steps of an I.C. process.  Demand for accurate process modelling is now high, both from technology development engineers who need it as a design tool to develop and optimise process technology, and from device development engineers who require accurate models of device structure as input to programmes modelling the electrical characteristics of the devices.  However, there exists at present a serious gap between the complexity of real processes now in development, and the capability of presently available process modelling programmes to simulate them.  This gap arises from shortcomings in all three essentials of a good process model viz (i) adequate physical theories which relate the process variables to the materials parameters involved, (ii) mathematical models that describe these relationships sufficiently accurately embodied in efficient and reliable computer routines that output the device structure in convenient form, (iii) relevant and accurate experimental data both as input to the model and as the ultimate check on the correspondence between the predicted and actual device structures.  The relationship between these three features is shown schematically in Fig.1(a).  The initial theory, model, and input parameters are used to simulate the separate sub-processes that together in sequence fabricate (hopefully!) the desired device structure. When a compatible set of sub-processes has been assembled into a provisional device process, actual structures are fabricated and the predictions of the process model at each sub process stage compared with experimental measurements of the partially fabricated device and test structures.  Discrepancies may indicate the need for more accurate basic parameter measurements (e.g. implant profiles, diffusion coefficients): or a more accurate or relevant

mathematical model or coding (e.g. finer mesh in numerical analysis): or
even a new physical theory of the process being modelled (e.g. effect of
chlorine ambients on diffusion coefficients in silicon). Improvements are
made in the modelling of the sub-processes until a simulated process when
actually fabricated yields working devices. Device measurements can then
be used as further experimental data to improve the simulation, and the
process optimisation phase begins. The crucial part that experimental data
plays in this is evident, and the emphasis in this paper will be on the
accuracy and relevance of available experimental data to process development, (Sections 3,4,5), what new data is needed to model VLSI processes
(Sections 6,7,8) and some experimental techniques required to obtain this
data, (Section 9). First a brief review of some existing process models
will be given.

## 2. Existing Process Modelling Programmes

Some of the existing process modelling programmes are shown in Fig.1b. All
the one-dimensional models listed give a graphical (and/or digital) output
of the calculated dopant (and/or carrier) concentration profiles along a
line perpendicular to the silicon surface. All surfaces, boundaries and
dopant concentration contours are taken to be parallel to the original
planar silicon surface and to extend to infinity in that plane. Three of
the programmes are freely available. The two-dimensional models also give
both graphical and digital outputs of calculated dopant and carrier concentrations, in the form of equi-concentration contours in a plane at right
angles both to the silicon surface and to the long boundaries of photo-
engraved structures in that surface. The long boundaries are taken to
extend to infinity. SUPRA, FEDSS and LADIS also model two-dimensional oxidation effects (e.g. selective oxidation) by semi-empirical methods of
varying flexibility. SAMPLE and SPUTTER are two-dimensional programmes
which model the parts of the structure outside the semiconductor: lithography of the photoresist and etching of the dielectric layer (SAMPLE); plasma
deposition and etching of metal layers on non-planar surfaces (SPUTTER).
Unfortunately, none of the 2D semiconductor process models is at present
available in this country. There is a strong and continuing interest in
the development and improvement of these and other process models as shown
by the number of recent meetings on the topic (SOGESTA 1982, LEUVEN 1983,
GALWAY 1983).

The most comprehensive process modelling programme which has been widely
used by many groups is the SUPREM II series, of which SUPREM II.5 is the
most recent version (1981). The main features of this are given in Fig.1(c).
The major operations in I.C. fabrication of the active device structure
(ion implantation, heat treatment, thermal oxidation, oxide deposition and
etching, and silicon epitaxy) are modelled. Sequences of these process
steps can be assembled in any order, and input and output require the minimum of specification by the user. The materials parameters (e.g. diffusion
coefficients, oxidation rates) can be specified by the user, but the mathematical models and algorithms cannot be changed. SUPREM II.5 has its own
set of materials parameters which are used if no user values are specified:
these default values are used in all the SUPREM simulations given in this
paper, except where otherwise stated. In the next few sections, simulations by SUPREM II.5 will be compared with experimental profiles and data,
to illustrate the extent to which a good currently available 1D process
simulator adequately models modern I.C. technology.

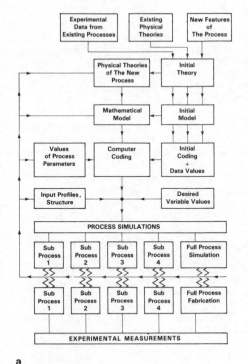

a

| Sub – Process | Effects Modelled | Features | |
|---|---|---|---|
| Ion Implantation | Dopant Profiles of Sb, As, B, P, in Si, SiO₂ | Sb, As, P  Half Gaussians | B  Pearson IV |
| Heat Treatment | Changes in above profiles caused by solid state diffusion and evaporation | Effect of high As, P concentration on diffusion coefficient | |
| Thermal Oxidation of Silicon | As above and growth of oxide and changes in profiles due to segregation, OED | Effects of oxidation on diffusion coefficient of boron | |
| Oxide Deposition | Addition of oxide layer with no other changes | | |
| Silicon Deposition by Epitaxy | As for heat treatment + growth of epi – layer and changes in profiles by autodoping | | |
| Oxide Etching | Complete or partial removal of oxide | | |
| Dopant Deposition | Constant Surface Concentration of B, P, As, Sb. | | |

Carrier concentration profiles and ($N_D$ – $N_A$) are also modelled in silicon region for all steps except ion implantation

Sequential Process

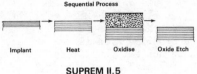

Implant　　　Heat　　　Oxidise　　　Oxide Etch

**SUPREM II.5**

c

| Name | Source | Features | Availability | Geometry |
|---|---|---|---|---|
| SUPREM II,5 | STANFORD USA | Sequential process model for dopant and carrier profiles and oxide thickness | Yes | 1D |
| SUPREM III | SILVAR-LISCO Belgium | More comprehensive and restructured version of SUPREM II,5 | ? | 1D |
| ICECREM | IFT Munich | Sequential process model similar to SUPREM II | Yes | 1D |
| MANELM | Southampton University | Coupled diffusion of As, B, Ga. | Yes | 1D |
| SUPRA | SILVAR-LISCO Belgium | Sequential process model for dopant and carrier profiles and oxide contours in two dimensions | ? | 2D |
| FEDSS | IBM USA | Sequential process model | No | 2D |
| LADIS | SIEMENS Munich | General diffusion model | No | 2D |
| LORD | BELL LABS. USA | General diffusion and oxidation model | No | 2D |
| TRIM | IFT Munich | Ion implanted range Distributions by Monte Carlo Methods | Yes | 2D |
| SAMPLE | BERKELEY USA | Sequential process model for lithography, etching, deposition | Yes | 2D |
| SPUTTER ETCH CS | PLESSEY Caswell | General plasma deposition and etch model | Yes | 2D |

b

Fig.1 Some features of integrated circuit process technology modelling programmes.
(a) General features of a good process model, showing the initial setting up of the provisional model, the simulation of the process steps (subprocesses) and the comparison with experimental measurements giving rise to an improved and modified model
(b) Some existing process modelling programmes
(c) Specific features of the Stanford programme SUPREM II.5

## 3. Ion implantation

Ion implantation is now universally used as the major doping technique in
I.C. technology, because of the accuracy and reproducibility of dopant dose
and spatial distribution, and the ease with which self-alignment techniques,
crucial for VLSI, can be implemented.  The energetic dopant ions come to
rest in the silicon with a concentration profile characteristic of that
dopant ion, dose, energy, implant angle and implant temperature.  The profile is very approximately Gaussian in shape; four examples are given in
Fig.2.  Those in Fig.2(a) are from data for implants into amorphous silicon
(Gibbons and Mylroie 1973) and show the characteristic skew towards the
surface for heavy ions (Sb) where nuclear stopping predominates (Maes 1981)
and towards the substrate for light ions (B) where electronic stopping predominates.  SUPREM II.5 uses two algorithms to accommodate the skew: for
As, P, and Sb it joins two half- Gaussians of different standard deviation;
for B it uses a Pearson IV (Hofker 1975) distribution.  In both cases, the
simulations fit the data quite well, apart from underestimating the width
of the antimony distribution by about 20%.  For implants into single crystal
silicon, the agreement is not as good (Fig.2b). The As simulation has a deeper peak, is narrower and is less skewed than the experimental profile,
(Hill et al 1976).  Moreover, the skew in both cases is in the opposite
sense to that expected from the amorphous target results.  The reason for
these discrepancies is the ability of ions in a single crystal to travel
without nuclear collisions in certain directions where open channels exist
characteristic of the crystal structure.  Although it is standard practice
to tilt silicon slices away from the primary beam (typically 7$^\circ$) to avoid
direct channelling, after several collisions, some ions will statistically
be in channelling paths, and so some tailing on all single crystal implants
is inevitable.  SUPREM II.5 does add an exponential tail to simulated boron
profiles, and this can be seen to approximate to the actual tail on the
main experimental boron profile; though the fit is rather fortuitous,
resulting from an underestimation of the peak depth and tail slope as compared with the experimental profile.  The differing tails observed on
slices where the slice rotation was not controlled indicate the occurrence
of additional planar channelling and illustrate both the difficulty of con-
trolling such tails experimentally and of simulating them accurately.  In
fact, none of the ion distribution theories accurately predicts the actual
distributions, (apart from Monte Carlo techniques which are very time con-
suming { Ishitani et al 1972 }).  In computer models, implant distribution
data is usually stored in terms of the four moments of a Pearson IV distribution, obtained by empirical fitting to experimental data.  The physical
theory plays little part in determining the mathematical model.

## 4. Diffusion

Solid state diffusion of dopants at high temperatures in silicon is often
made use of to redistribute dopants to the desired final distribution.
However, diffusion is also often an unwanted by-product of heat treatments
required for other reasons, e.g. anneal of implantation damage, thermal
oxidation.  In either case, the capability to model the redistributions
occurring at each process stage accurately is essential if the final device
structure is to be adequately represented.  At low concentrations and under
true equilibrium conditions diffusion behaviour can be characterised by an
intrinsic diffusion coefficient for each dopant which only depends on temperature:

$$D_i = D_o \exp - {}^Q/_{KT} \qquad \ldots \qquad \ldots \qquad \ldots \qquad (1)$$

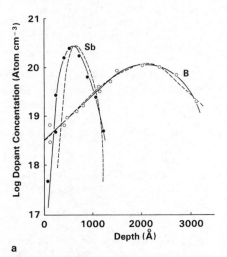

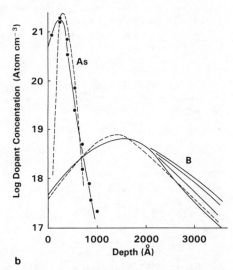

a                                                        b

Fig.2  Experimental Ion Implanted Dopant Profiles in Silicon compared with
SUPREM Simulation (dotted lines)  (a) Amorphous Silicon Matrix, data from
Gibbons and Mylroie 1973  (b) Crystalline Silicon Matrix, data from Hill and
Chater 1984.  Main boron profile implanted at 7° tilt, 15° rotation: other
boron implants at 7° tilt, uncontrolled rotation.

| DOPANT | REFERENCE | $D_o$ cm$^2$sec$^{-1}$ | Q eV | 900°C | 1000°C | 1100°C | 1200°C |
|---|---|---|---|---|---|---|---|
| | | | | DIFFUSION COEFFICIENT cm$^2$sec$^{-1}$ | | | |
| | HILL (1981) | 24 | 3.87 | 5.7E-16 | 1.2E-14 | 1.5E-13 | 1.4E-12 |
| B | OKAMURA (1969) | 5.1 | 3.70 | 6.5E-16* | 1.2E-14* | 1.4E-13 | 1.1@-12 |
| | LIN et al (1981) | | | 1.5E-15 | 1.5E-14 | 1.8E-13 | 1.7E-12 |
| | SUPREM II.5 | | | 1.1E-15 | 1.5E-14 | 1.5E-13 | 1.1E-12 |
| | HILL (1981 | 0.6 | 3.51 | 5.0E-16 | 7.7E-15 | 7.9E-14 | 5.9E-13 |
| P | MAKRIS & MASTERS(1973) | 5.3 | 3.69 | 7.5E-16* | 1.3E-14 | 1.5E-13 | 1.3E-12 |
| | LIN et al (1981) | | | 6.7E-16 | 1.3E-14 | 1.4E-13 | 1.1E-12 |
| | SUPREM II.5 | | | 7.3E-16 | 1.3E-14 | 1.4E-13 | 1.2E-12 |
| | HILL (1981) | 13 | 4.05 | 5.2E-17 | 1.2E-15 | 1.8E-14 | 1.8E-13 |
| As | MASTERS & FAIRFIELD(1969) | 60 | 4.20 | 5.5E-17* | 1.4E-15 | 2.3E-14 | 2.6E-13* |
| | CHIU & GHOSH(1971) | 24 | 4.08 | 7.2E-17* | 1.7E-15 | 2.6E-14 | 2.7E-13 |
| | SUPREM II.5 | | | 5.50E-17 | 1.34E-15 | 2.05E-14 | 2.15E-13 |

TABLE 1 - Intrinsic diffusion coefficients of boron, phosphorus and
          arsenic.  Values marked with an asterisk are extrapolated
          outside the experimental range studied.

| DOPANT | REFERENCE | 900°C | 1000°C | 1100°C | 1200°C |
|---|---|---|---|---|---|
| | | DIFFUSION COEFFICIENT cm$^2$sec$^{-1}$ | | | |
| B | HILL (1981) | 4.2E-15 | 3.5E-14 | 2.6E-13 | 1.8E-12 |
| | SUPREM II.5 | 6.1E-15 | 3.4E-14 | 2.1E-13 | 1.2E-12 |
| P | HILL (1981) | 2.5E-15 | 2.1E-14 | 1.4E-13 | 8.4E-13 |
| | SUPREM II.5 | 1.3E-15 | 2.3E-14 | 2.6E-13 | 2.1E-12 |
| As | HILL (1981) | 2.2E-16 | 2.2E-15 | 2.3E-14 | 2.0E-13 |
| | SUPREM II.5 | 5.5E-17 | 1.3E-15 | 2.0E-14 | 2.1E-13 |

TABLE 2 - Comparison of SUPREM default values and experimental values
          of boron diffusion coefficient in (100) silicon in dry
          oxygen ambient.

Experimental values for $D_i$, for B, P, and As are compared with the SUPREM default values in Table 1; the agreement is good, SUPREM values always lying within the experimental spread (maximum spread ±20%, maximum $\sigma$ ±11%, mean of $\sigma$ values = ±5%) which represents only about ±4% uncertainty in spatial position $(N\sqrt{Dt})$ at any point on the profile.

Intrinsic equilibrium conditions seldom occur in real processing, and often oxidation-modified diffusion behaviour is more relevant. This can be represented for B, P and As (Hill 1981) as

$$D_{ox} = D_i + \Delta D_{ox} \qquad \ldots \quad \ldots \quad \ldots \quad (2)$$

$$\Delta D_{ox} = \Delta D_o \exp \frac{-\Delta Q}{KT} \qquad \ldots \quad \ldots \quad \ldots \quad (3)$$

SUPREM uses these algorithms for boron diffusion only; for phosphorus, under oxidising conditions $D_o$ is increased by a constant factor. Experimental values of $\Delta D_{ox}$ and $\Delta Q$ are given in Fig.3(d) for various oxidising ambients and slice crystallographic orientations (Hill 1981). In Table 2 some of these experimental values (for boron in {100} silicon in dry Oxygen) are compared with SUPREM simulation values. In this case, agreement is worse, SUPREM values differing by up to 45% and mean of the fractional deviations being 25%. SUPREM does not have different default values for diffusion in (110) silicon, or with water-containing ambients. It is evident from the graphical plots of this experimental data, Figs.(3a & 3b) that diffusion coefficients are generally increased during oxidation, i.e. $\Delta D$ is positive, and SUPREM only allows for this case. However, at high temperatures or in chlorine-containing ambients, $\Delta D$ can be negative. This change-over occurs at lower temperatures in (111) material than in (100), as seen by the increase in D in the order $D_{ox}(111) \rightarrow D_i \rightarrow D_{ox}(100)$ in the experimental data of Fig. 3(c). Since there is a direct correlation between the occurrence of negative $\Delta D$ and the desirable processing condition of shrinkage of stacking fault defects (Hill 1980, Claeys et al 1981) adequate modelling of such processing conditions is needed. This will be difficult, since in this regime, time-dependent and depth dependent diffusion behaviour occurs because of the similar concentrations of the mutually annihilating populations of interstitial and vacancy defects responsible for the oxidation-modified diffusion (Gösele and Tan 1983).

At high dopant concentrations, diffusion behaviour deviates even more strongly from intrinsic behaviour. Typical experimental profiles for As, P and B under these conditions are shown in Fig.4. For As and P, the theory of concentration-enhanced diffusion is reasonably well understood, (Chiu & Ghosh 1971, Fair and Tsai 1977) based on the change in concentration of singly and doubled charged (-ve) vacancies with the free carrier concentration. Algorithms based on these models are used in SUPREM. For arsenic, SUPREM gives the approximate shape of the diffused distribution, but in order to obtain the fit shown in Fig.4(a), $D_o$ had to be increased by X 1.53. The diffusion tail on phosphorous profiles (Fig.4b) is reasonably well-fitted by SUPREM. Boron diffusion at high concentrations is only now being seriously studied, and shows both a diffusion tail and an immobile peak as shown in Fig.4c (Godfrey et al 1983). SUPREM does not model these effects. The same reference reports changes in the intrinsic coefficient for boron by a factor of 6 in the range $10^{16} - 10^{19}$ atoms/cc; no mechanism is known, but the effect will certainly require modelling for CMOS structures.

When dopants at high concentrations are in proximity to other dopants, additional modification of the diffusion behaviour of each can occur.

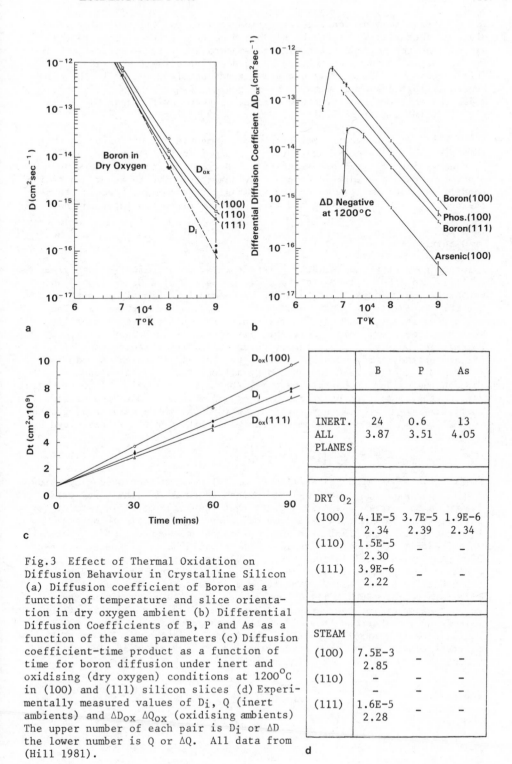

Fig.3 Effect of Thermal Oxidation on Diffusion Behaviour in Crystalline Silicon (a) Diffusion coefficient of Boron as a function of temperature and slice orientation in dry oxygen ambient (b) Differential Diffusion Coefficients of B, P and As as a function of the same parameters (c) Diffusion coefficient-time product as a function of time for boron diffusion under inert and oxidising (dry oxygen) conditions at 1200°C in (100) and (111) silicon slices (d) Experimentally measured values of $D_i$, Q (inert ambients) and $\Delta D_{ox}$ $\Delta Q_{ox}$ (oxidising ambients) The upper number of each pair is $D_i$ or $\Delta D$ the lower number is Q or $\Delta Q$. All data from (Hill 1981).

|  | B | P | As |
|---|---|---|---|
| INERT. ALL PLANES | 24 3.87 | 0.6 3.51 | 13 4.05 |
| DRY $O_2$ |  |  |  |
| (100) | 4.1E-5 2.34 | 3.7E-5 2.39 | 1.9E-6 2.34 |
| (110) | 1.5E-5 2.30 | – | – |
| (111) | 3.9E-6 2.22 | – | – |
| STEAM |  |  |  |
| (100) | 7.5E-3 2.85 | – | – |
| (110) | – | – | – |
| (111) | 1.6E-5 2.28 | – | – |

These interactions are particularly important in modelling bipolar transistor structures, where the highly doped emitter is very close to the narrow base region, and the device parameters are very sensitive to basewidth. Dopant-dopant interactions have been intensively studied (Willoughby 1977, Fair 1981) but are not adequately modelled in the present sequential process programmes. The programme MANELM does satisfactorily model the emitter-base structure itself, as shown by the results in Fig.4(d), (Mallam et al 1981).

The overall situation with existing process models for diffusion is that they are a useful approximate guide to real diffusion behaviour, and if checked against a few experimental results from the actual processing situation, can be used as a quantitative extrapolation to other processing conditions. The models require considerable development, however, to meet the needs of VLSI processing and these will be discussed later in that context.

## 5. Oxidation

A lot of good experimental data exists for the basic oxidation behaviour of (111) and (100) silicon in dry oxygen, e.g. (Deal & Grove 1965), and wet oxygen, e.g. (Deal 1978) and SUPREM models this behaviour quite well for thicker oxides (greater than 200Å) and atmospheric pressure. The model used is not, however, fully adequate for thin oxides, especially when non-atmospheric dry $O_2$ pressure exists. This is because of shortcomings in the Deal & Grove theory which postulates three regions of oxidation behaviour according to oxide thickness: (i) thick oxides (>5000Å), oxidation rate controlled by diffusion of oxygen molecules through the oxide, oxide thickness $X \underset{\sim}{=} \sqrt{Bt}$  (ii) thin oxides (>200Å), oxidation rate controlled by reaction of oxygen molecules with the silicon surface, $X \simeq \frac{B}{A} t$  (iii) very thin oxides (0-200Å) where very rapid growth to 200Å in $A$ time $\tau$ is assumed. The whole oxide growth is described by

$$X = {}^A/_2 \left( (1+\{t+\tau\}\frac{4B}{A^2})^{\frac{1}{2}} - 1 \right) \quad \ldots \quad \ldots \quad \ldots \quad (4)$$

SUPREM gives some crude capability of modelling very thin oxides by using X 10 linear rate constant $(^B/_A)$ for oxides below 200Å. The thick oxide regime is satisfactory and it appears that the basic model is correct for this regime. In the thin oxide regime, however, the basic theory is incorrect, since it predicts a linear pressure dependence of $^B/_A$, whereas experimentally (van der Meulen 1972, Lie et al 1982) the exponent is near 0.7. A revised model (Blanc 1978) postulating a rate controlled by the reaction of oxygen atoms with the silicon surface predicts a pressure exponent of 0.5, and allows a better fit to the data. This model is incorporated into SUPREM III. However, an improved model giving the correct exponent for the pressure dependence of $^B/_A$ and the correct logarithmic growth (Massoud et al 1982) has been proposed (Hu 1983): the oxidation rate is controlled by chemisorption of oxygen molecules onto the silicon surface in the thin oxide case, and by transport of electrons and $O^-$ ions in very thin oxides. The growth is described by :

$$X = 1/_{2a} (p^{-m} - p^{1-m}) \quad 1 \geqslant p \geqslant 0 \quad \ldots \quad \ldots \quad (5)$$

and

$$t = -\frac{1}{2} \int_1^p \frac{mp^{-m-1} + (1-m)p^{-m}}{bp^m + aC_2 exp\left[-(p^{-m}-p^{1-m})/2aL_2\right]} dp \quad (6)$$

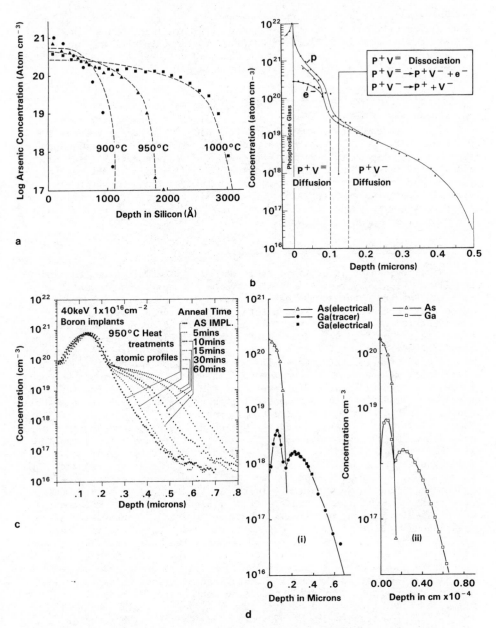

Fig.4  Effects of High Dopant Concentration on Diffusion Behaviour in
Crystalline Silicon (a) Arsenic Profiles after Ion Implant and Anneal
(30KeV, 5E15 dose + 30 min in Argon); experimental data from (Hill et al
1976): SUPREM simulation uses 1.53 x default value of $D_O$ (b) Phosphorus
profile, diffusion from chemical source (100 ppm $PH_3$, 1000ppm $O_2$ for 60
mins at 850°C) showing regions of different diffusion behaviour (Smith and
Hill 1971) (c) Boron Profiles after Implant and Anneal from (Godfrey et al
1983) (d) As + Ga Profiles after simultaneous diffusion at 1000°C showing
(i) experimental profiles (ii) MANELM simulation (Mallam et al 1981).

The dependence of the linear rate constant $B/A$ on silicon crystal orientation and on silicon doping (Ho & Plummer 1979) are incorporated into SUPREM II, but not the effects of chlorine in the oxidising ambient (e.g. Deal 1978). Satisfactory fits can be obtained by altering the default values for the linear and parabolic rate constants.

## 6. Modelling Requirements of VLSI

The demand for very large scale integrated circuits necessarily means individual circuit components must shrink laterally. The device physics requires that a vertical shrinkage occurs too, though not linearly, nor the same for all components. Higher speed devices also require smaller dimensions to minimise transit times of carriers, and in bipolar devices this is a strong additional motive for vertical shrinkage. The change in vertical dimensions of bipolar transistors is well illustrated by Fig.5, which shows structures now in production, in development and in the research stage. The overall shrinkage is a factor of 10, from 7000Å to 700Å.

## 7. Vertical Shrinkage of Device Structures

Such drastic changes in vertical scale are creating two types of demands on modelling (i) higher resolution data for the shallower dopant profiles used (ii) new data relevant to the new processing technologies which are being introduced to fabricate such structures (e.g. molecular implants, pre-amorphisation of substrates, polysilicon structures, transient anneals, high pressure oxidation). The need for shorter times to anneal implantation damage is evident from the time-temperature map in Fig.6a. This compilation of experimental data (Hill 1983) shows the minimum time-temperature cycle for full anneal of two common process implants superimposed on contours of constant dopant redistribution (high concentration arsenic, $x = 3\sqrt{Dt}$). Today's technology where 1 micron redistribution is acceptable can be effected with furnaces and times greater than 30 minutes. VLSI demands less than 0.1 micron redistribution and hence transient anneals of 5-50 secs. Diffusion behaviour at such short times cannot be extrapolated from existing data, as Fig.6b shows. The diffusion coefficient is about 2.5 x greater than predicted. Much larger discrepancies are reported (Fair et al 1983), and the influence of non-equilibrium point defects generated by damage annealing is inferred. In fact, VLSI technology needs models that treat non-equilibrium and kinetics of reaction of point defects and dopant species in many areas: e.g. atomic clustering, dopant segregation, grain boundary diffusion, time-dependent diffusion coefficients, anneal of implant damage. An example of the discrepancy that can occur between simulation and experiment when the kinetics of As clustering are not taken into account is shown in Fig.7 (Scovell 1980). Another consequence of vertical dimension shrinkage is the higher dopant concentrations in some parts of devices and for these, existing models do not predict electrical properties well. A recent compilation of carrier mobility for highly doped silicon (Hill 1982) is given in Fig.8. The electron mobility $\mu_n$ behaves quite differently for $n > 10^{20}$/cc than expected from lower concentration data.

$$n > 10^{20}/cc : \mu_n = 7.6 \times 10^{+11} n^{-\frac{1}{2}} \quad \ldots \quad \ldots \quad \ldots (7)$$

## 8. Lateral Shrinkage of Device Structures

The changes in device structure and thus in fabrication technology forced by the trend to smaller lateral geometries are resulting in much less planar structures than previously. This is clear when present and future MOS structures are compared, as shown in Fig.9. The schematic MOS transistor

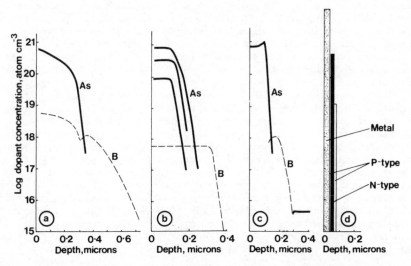

Fig.5 Vertical shrinkage of bipolar transistor structures to achieve
higher frequency operation. Vertical dopant concentration profiles
through (a) present production transistor (b) laser processed structure
(c) SIPOS polysilicon emitter structure (d) hot electron transistor
(Hill 1983).

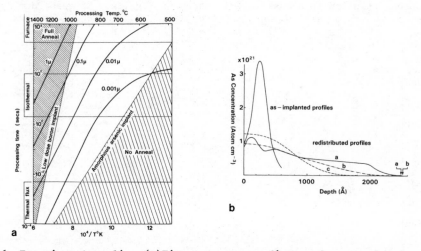

Fig.6 Transient Annealing (a)Time-temperature diagram for anneal and
redistribution of dopants by solid-state mechanisms. The straight lines
represent the minimum time-temperature schedules for full anneal of (i) a
40KeV low dose boron implant and (ii) a 40KeV amorphising arsenic implant
based on published experimental values. The family of curves gives the
loci of time-temperature schedules for redistribution $(3\sqrt{Dt})$ of a $10^{16}$
ions/cm 40KeV arsenic implant by 1, 0.1, 0.01 and 0.001 micron. From (Hill
1983). (b)Redistribution of Arsenic by Transient Annealing. Implant 30KeV
$10^{16}$ ions/sq.cm, Anneal 25 secs at 1100°C in air (A.G."Heatpulse" Annealer).
Solid lines are experimental profiles (Hill and Skinner 1984), dotted lines
are SUPREM simulations using 'c' default values, 'b' 3 x default value of
$D_0$. Arrows indicate junction depths at $10^{17}$ atom $cm^{-3}$.

in Fig.9b has very little area that could be modelled as a planar structure. In particular, both the isolation techniques used, illustrated in Fig.10a, expose a number of crystal orientations to the processing ambient. Modelling the dopant redistributions in such structures during subsequent oxidations is complex, since the enhancement of diffusion coefficients under such conditions (Section 4) has been shown to be determined by the crystal orientation of the nearby oxidising plane, as shown in Fig.10b (Hill 1981). The single diffusion coefficient of the planar structure is thus replaced by a 2 dimensional distribution of diffusion coefficients. Small lateral dimensions also mean that other two-dimensional effects become important in VLSI modelling. The edge portions of devices become a greater fraction of the whole, and device behaviour is more influenced by the non-planar distributions of dopants under mask edges (Fig.11a). Ion implanted dopant distributions in these regions are sensitive to mask edge profile (Fig.11b) and to the ion beam direction, which is necessarily not 90° in electrostatic implanters or where ion channelling is to be avoided (Fig.11c). In selective oxidation, mask edge effects can dominate the rate of oxidation even in the planar region of very small structures (Fig.11d), through the increasing influence of hydrostatic stress as device size decreases.

Modelling of stress will be increasingly important, since it affects the device structure in many ways. Tensile stress decreases the band-gap of silicon, and hence reduces diffusion coefficients through lattice misfit stress at high dopant concentrations, or through external stresses at all dopant concentrations, (Fair 1981). The effect of deposited oxide stress on boron diffusion at a mask edge is shown in Fig.12a. Stresses during local oxidation of silicon change the oxygen diffusion behaviour, thus

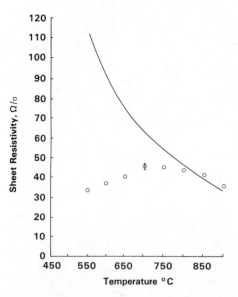

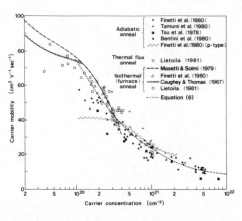

Fig.7 Electrical Properties of Arsenic Implant (150KeV, 6E15 ions/ sq.cm.) after Anneal (30 mins). Curve = SUPREM Simulation, Points are Experimental(Scovell Young 1980)

Fig.8 Electrical Properties of Heavily Doped Silicon. Electron mobility as a function of carrier density in n-type silicon: hole mobility in p-type silicon shown by wavy line. Line labelled Eqn.6 refers to equation 7 in this paper. References are to original source (Hill 1982).

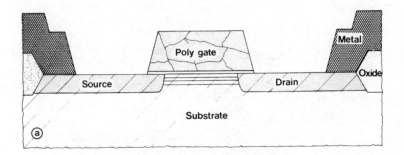

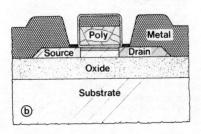

Fig.9  Lateral Shrinkage of MOS transistor structures to achieve
VLSI component densities.  Vertical sections through (a) typical
transistor fabricated on 2½ micron process  (b) potential transistor fabricated on a 1 micron silicon-on-insulator process
(Hill 1983).

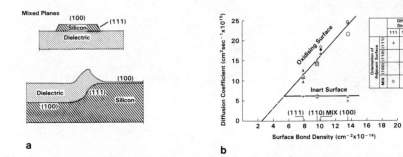

Fig.10  Effect of Non-Planar Topography on Diffusion Behaviour in
Crystalline Silicon (a) non-planar topographies generated by the mesa-
etching and selective oxidation isolation technologies.  The crystal
planes exposed to subsequent oxidation are indicated  (b) Diffusion
coefficient of boron at 977°C as a function of the crystallographic
orientation of the adjacent oxidising surface (Hill 1981).

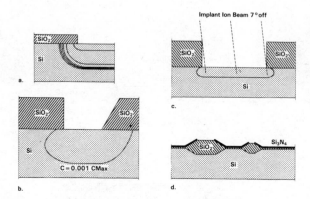

Fig.11  Two-dimensional effects of lateral shrinkage on process modelling
(a) Increasing importance of the non-planar dopant distributions at component edges: dopant concentration contours at a diffused window edge
(b) Effect of mask edge profile on implanted dopant concentration contour
(Hill 1980) (c) Asymmetry caused by non-vertical ion beams   (d) Effect of
nitride stress on oxidation rate in 1.5 and 0.5 micron windows (Itsumi et
al 1983)

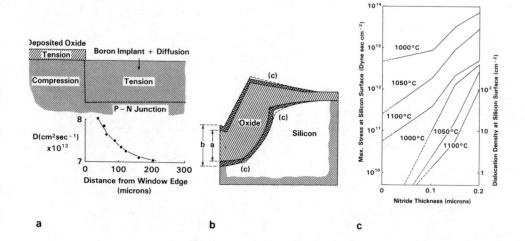

Fig.12  Effects of mechanical stress in silicon processing (a) Diffusion
coefficient of boron under stress from deposited oxide mask: calculated
values, curve (Fair 1981), experimental values, points (Todokoro et al
1978). (b) Selective oxidation of silicon under self-induced hydrostatic
stress: simulation of oxide boundaries "a" after 33.5 mins, "b" after
57.5 mins, "c" after 57.5 minutes without stress in model.  After
(Poncet 1983). (c) Selective oxidation under nitride stress; calculated
stress and measured dislocation densities.  After Chin et al (1982).

altering the "birds beak" morphology (Fig.12b), (Poncet 1983).  At higher
stresses, dislocations are formed and the onset of this undesirable side
effect can be predicted if the stress at the silicon surface is modelled
(Chin et al 1982), as shown in Fig.12c.  Stress also has a direct effect on
the electrical behaviour of devices, as shown by the change in mobility of
MOS devices in silicon-on-insulator technology, where stress can be changed
from tensile through neutral to compressive (Tsaur et al 1982).  It is
important that stress distributions should be modelled right through the
device fabrication process, since defect-forming local stresses can often
result from transient combinations of stresses which are not evident from
the final device structure.

## 9. Experimental Techniques

The great interest in developing computer models for many of the foregoing
effects (SOGESTA 1982, LEUVEN 1983), has revealed the inadequacy of presently available experimental data to check the models against, as described in Section 1.  This has stimulated the improvement of existing
experimental techniques and the development of some new ones, and new
experimental data is becoming available.  The higher resolution one-
dimensional dopant profiles required to characterise the shallower layers
created by the vertical shrinking of structures are mainly being obtained
by improvements to existing profiling techniques.  Most progress has been
made in optimising sputter-depth profiling methods for microelectronic
structures (Zinner 1983), in which planar layers are removed from the semiconductor surface by continuously sputtering away the surface atoms with a
well-controlled ion beam, and monitoring the composition of either the
emitted ions (e.g. SIMS technique) or the new surface (e.g. Auger technique).  In the past the depth resolution of these techniques was limited
by profile broadening due to atomic mixing, knock-on of ions and non-
planar removal of material, and it has long been realised (Schulz et al
1973) that these effects are minimised by using low energy ion beams at
small angles of incidence.  This, however, also reduces the sensitivity of
the ion-collection techniques, and most effort in the past has been to
optimise sensitivity.  Now, many experimental systems are being optimised
for depth resolution and better spatial resolution is available.  The
effect of optimising ion-energy and incidence angle can be seen in  Fig.13a
which shows oxygen profiles through an $SiO_2$ layer sandwiched between a
polysilicon layer and the silicon substrate.  As beam energy and incidence
angle are reduced, the oxygen profile approaches the step-function expected
from TEM measurements of the oxide thickness ( 58Å ).   Currently, much
effort worldwide is being directed towards obtaining this high spatial
resolution for SIMS while retaining the high dopant sensitivity, (Williams
and Tsong 1983).  One important feature of both SIMS and Auger is that a
number of dopants can be profiled simultaneously, so that the relative
accuracy of the profiles is far higher than the absolute accuracy.  An
example is shown in Fig.13(b), where the effect of an oxide layer at the
polysilicon-silicon interface on the diffusion of arsenic through the
interface is monitored by Auger analysis.  It can be clearly seen that the
58Å oxide blocks the arsenic diffusion, whereas the 12Å oxide does not,
with a relative accuracy of ±10Å.  Auger analysis can also be used to
establish ion-implantation profile data far more accurately in the peak
region of high dose implants, as shown in Fig.13(c).  On this fine scale,
the SUPREM data is seriously in error as to peak depth, profile width and
profile shape.  A third technique, important because of the absolute nature
of the depth and concentration calibration, the non-removal of material,
and the extra information on the damage profile, is Rutherford Back

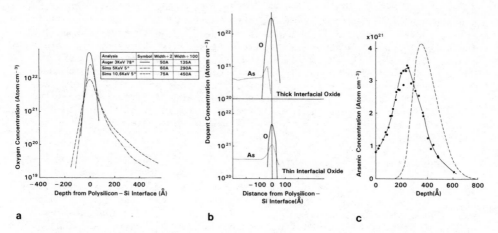

Fig.13  High resolution Dopant Profiling techniques   (a) Oxygen Concentration Profiles through a Polysilicon-Oxide-Silicon Structure: sputtering beam angle and energy as shown.  Data from SIMS (Hill & Littlewood 1984), Auger (Hill & Skinner 1984).   (b) Oxygen and Arsenic Concentration  Auger Profiles through Poly-SiO$_2$-Si structures.  Oxide thicknesses are 58Å and 12Å (Hill & Skinner 1984).   (c) Arsenic Implant Profile (40KeV, $10^{16}$ ions/cc): experimental points from (Hill & Skinner 1984), dotted curve = SUPREM simulation.

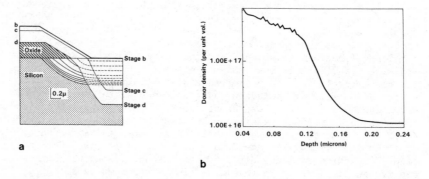

Fig.14  High Resolution Carrier Profiling Techniques   (a) Two-dimensional technique based on anodic sectioning of the non-planar portion of a resistor: experimentally determined evolution of the section after "b" implant and diffusion, "c" plasma etching, "d" 17 sequential anodic removals of silicon layers.   (b) Bipolar transistor base doping profile derived from C-V measurements on completed devices (Daniels 1983).

Scattering. Specifying the damage profile at various process steps will be important because of its effect on diffusion and lattice stress (Fair 1981), and the electrical effects of remaining damage in the device. Severe damage ($>10^{19}$ displaced atoms/cc) can be profiled by RBS (Hemment et al 1983), lighter electrically active damage ($>10^{13}$ centres/cc) can be profiled by DLTS techniques (Troxell 1983).

Electrical measurements can be used to establish dopant profiles if the ionisation, compensation and carrier mobilities are independently known or determined. In some cases, dopant profiles in the final device structure can be determined from electrical parameters. An example is given in Fig.14a, where careful analysis of capacitance-voltage measurements on a bipolar device structure yields the base profile ($N_A$-$N_D$) shown (Daniels 1983). The long-established anodic stripping and sheet resistance technique to establish 1D carrier profiles (Shaw 1973) is currently being developed at Caswell as a high spatial resolution two-dimensional profiling technique. The non-planar part of resistors is sectioned anodically, and the resistivity-distance data is computer-processed to generate the carrier concentration contours at the resistor edges. The evolution of the vertical edge during one experiment is shown in Fig.14b.

Transverse transmission electron microscopy will play an essential role in establishing the actual topography and internal structure of VLSI devices. Much of this information has been obtained by scanning electron microscopy in the past: most theories of "birds beak" formation are based on such measurements (e.g. Bassous et al 1976). Much higher resolution is required for VLSI, and superb quantitative measurements are becoming available, as shown in Fig.15(a) (Vanhellemont et al 1983). Some information on dopant distribution can also be obtained by developments of transverse TEM, as can be seen from the etch-delineated junction contour in Fig.15(b) (Yallup et al 1983). Even higher resolution than these is required to characterise some structures, however, and lattice imaged TEM is becoming important for device structure analysis. An example is shown in Fig.15(c) where the interface region of a high-gain polysilicon-silicon bipolar transistor has been sectioned. One theory postulates a very thin layer of oxide as a tunnelling barrier (de Graeff and de Groot 1979) and at this resolution, it can be seen that this particular device has indeed a continuous amorphous phase 12Å thick at the interface.

## 10. Summary

The accuracy of process modelling programmes for present and future VLSI device structure has been shown to be critically dependent on the availability of relevant, precise and comprehensive experimental data. Present process models, used with care, can give sufficiently accurate simulations of planar structures to considerably reduce time in setting up provisional process schedules. More accurate simulations result from feeding back experimental process data into the model, but the algorithms for implantation, diffusion and oxidation are not sufficiently sophisticated to allow exact modelling. Simulation accuracy will be considerably worse in VLSI device modelling, where the complexities of high concentrations, new materials, non-planar topography, transient heat treatments, and two dimensional modelling of dopant, stress and damage profiles will be added to those of existing processes. New process models and experimental data are urgently required, and this need has stimulated new work in physical theories, numerical techniques and high resolution experimental techniques of profile and structure measurement; much more work is, however, required.

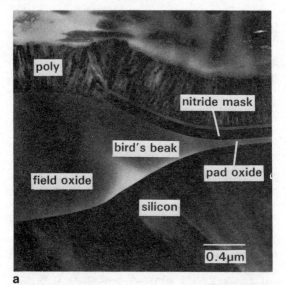

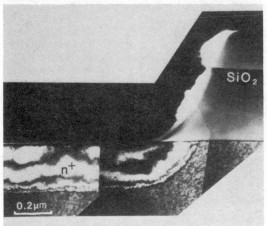

Fig.15   Transverse TEM Techniques for Two-Dimensional Data on Device Structure and Doping.

(a)   TTEM section through the "birds beak region" of a selectively oxidised device (Vanhellemont et al 1983) showing the thin (600Å) "pad" oxide, the 900Å nitride overlayer forced upwards by the lateral penetration of the thick oxide, and the thin oxide formed from the oxidation of about 100Å of nitride.   The columnar layer is deposited polysilicon to protect the sample during sectioning.

(b)   TTEM section through the edge of an oxide-masked implant window, after a 40KeV, $1\times10^{16}$ ions/cm$^2$ arsenic implant annealed 60 mins at 950$^o$C, showing junction contour delineated by selective etching (Yallup et al 1983).

(c)   Latticed imaged high resolution TTEM of polysilicon-silicon dioxide-silicon substrate structure (reading from top to bottom) showing the crystalline nature of the silicon regions, and the amorphous nature of the 12Å silicon dioxide layer.   (Hill and Vanhellemont 1984 to be published).

## 11.Acknowledgements

We thank all our colleagues for their support in preparing this paper, particularly our collaborators, John Kilner, Richard Chater and Steve Littlewood (Imperial College), Jan van Hellemont (Antwerp Univ.), Derek Skinner (Caswell) for permission to reproduce figures from our unpublished work: and to Derek Godfrey and Kevin Yallup (GEC Hirst), Peter Scovell (STL) and Arthur Willoughby (Southampton Univ.) for permission to reproduce published data.

Some of this work was carried out with the support of the Procurement Executive MOD, sponsored by DCVD, and thanks are due to them and Plessey Research (Caswell) Ltd. for permission to publish.

## 12.References

Blanc J 1978 Appl. Phys. Letts. 33 424
Bassous E, Yu H N, Maniscalco V 1976 J. Electrochem. Soc. 123 No.11 1729
Chiu T C and Ghosh H N (1971) IBM J. Res. Dev. Nov. 472
Chin D, Oh S Y, Hu S M, Dutton R W, Moll J L (1982) IEDM Technical Digest 1982 IEEE Piscataway N.J. 228
Claeys C L, Laes E E, Declerck G J, van Overstraeten R J 1977 Semiconductor Silicon 1977 (Huff H R & Sirtl E eds) J. Electrochem. Soc. Princeton N.J. 773
Deal B E and Grove A S (1965) J. Appl. Phys. 36 3770
Deal B E 1978 J. Electrochem. Soc. 125 576
Daniels I 1983 This Conference Paper S12.2
de Graaff and de Groot 1979 IEEE Trans. Electron. Dev. ED26 1771
Fair R B, Wortman J J, Liu J, Tischler M, Masnari N A, Duh K Y 1983 Paper IIIB-8 41st Device Res. Conf. June 20-22 University of Vermont, USA
Fair R B and Tsai J C C (1977) J. Electrochem. Soc. 124 1107
Fair R B (1981) In. Materials Processing Theory and Practices (Wang F Y Y ed) North Holland New York 315
GALWAY 1983 NASECODE III Conference: Finite Element Programming with Special Emphasis on Semiconductor Device and Process Modelling 13-14 June Galway Ireland
Gibbons J F and Mylroie S 1973 Appl. Phys Letts. 22 568
Gösele U and Tan T Y 1983 In Defects in Semiconductors (Mahajan S and Corbett J W eds) North Holland New York
Godfrey D J, Groves R D, Willoughby A F, Dowsett M G, 1983. This Conference Paper S11.5
Hemment P L F, Maydell-Ondrusz E A, Stevens K G 1983 Electron. Letts. (GB) 19 No.13 483
Hofker W K 1975 Rad. Eff. 24 223 and 25 205
Hill C, Boys D R, Dearnley G, Wilkins M A (1976) Paper 9 IOP Conf. on Materials and Processing Effects in Semiconductor Devices, Brighton March 8-10
Hill C 1980 Device Impact of New Microfabrication Technologies Course Notes Vol.III, University Leuven Belgium
Hill C 1981 In Semiconductor Silicon (1981) (Huff H R, Kriegler R J, Takeishi Y eds) Electrochem. Soc. Princeton N.J. 988
Hill C 1982 Chap. 13 In Laser & Electron Beam Processing of Semiconductor Structures (J M Poate & J W Mayer eds) Academic Press New York 479
Hill C 1983 In Laser Solid Interactions and Transient Thermal Processing of Materials (Naryan J, Brown W L, Lemons R A eds) North Holland New York 381

Hill C and Chater R 1984  To be published
Hill C and Littlewood S 1984 To be published
Hill C and Skinner D 1984  To be published
Ho C P and Plummer J D 1979  J. Electrochem. Soc. 126 1516
Hu S M 1983 Appl. Phys. Letts. 42 (10) 872
Ishitani T, Shimizu R, Murata K, 1972 Phys. Stat. Solidi B50 681
Lin A M R, Antoniadis D A, Dutton R W 1981  J. Electrochem. Soc. 128 1131
LEUVEN 1983: VLSI Process and Device Modelling: Course Notes University
  of Leuven  Belgium
Lie L, Razouk R, Deal B 1982  J. Electrochem. Soc. 129 2828
Maes H, Vandervorst W, van Overstraeten R 1981 In Impurity Doping Pro-
  cesses in Silicon (Wang F F Y  ed) North Holland  Amsterdam 443
Makris J S and Masters B J 1973  J. Electrochem. Soc. 120 1252
Mallam N, Jones C L, Willoughby A F W 1981 In Semiconductor Silicon 1981
  (Huff H R, Kriegler R J, Takeishi Y  eds) Electrochem. Soc. Pennington
  N.J. 979
Massoud H Z, Ho C P, Plummer J D 1982 Stanford Univ. Tech. Report
  TRDXG501-82
Masters B J and Fairfield J M 1969  J. Appl. Phys. 40 2390
Okamura M 1969  Jap. J. Appl. Phys. 8 1440
Poncet A 1983: Finite Element Simulation of Local Oxidation of Silicon
  LEUVEN 1983
SOGESTA 1982: NATO Study on Process and Device Simulation for MOS-VLSI
  Circuits Urbino Italy July 12-23
Scovell P D and Young J M 1980 Electronics Letts. 16 (16) 614
Schulz F, Wittmack K, Maul J 1973 Rad. Eff. 18 211
Shaw D 1977 Atomic Diffusion in Semiconductors Plenum Press London
Smith R S and Hill C 1971  Unpublished data
Todokoro Y and Teramoto I 1978 J. Appl. Phys. 49 3527
Tsaur B-Y, Fan J C C, Geis M W 1982 Appl. Phys. Letts. 40 (4) 322
van der Meulen Y J 1972  J. Electrochem. Soc. 119 530
Vanhellemont J, Claeys C, van Landuyt J, Declerck G, Amelinckx S,
  van Overstraeten R 1983 In Microscopy of Semiconducting Materials (1983)
  (Cullis A G, Davidson S M, Booker G R  eds) IOP London
Williams P and Tsong I S T 1983 Proceedings of 6th International Conf. on
  Ion Beam Analysis Proc. Vol. Nucl. Inst. and Methods
Willoughby A F W  1977  J. Phys. D. Appl. Phys. 10 455
Zinner E  1983  J. Electrochem. Soc. 130 No.5  199C